Purpose

The purpose of this investigation was to offer recommendations with respect to (1) whether Afghanistan should make use of television for its educational activities; and if it should (2) which of the methods of utilizing television, adopted in developing countries, would appear to be the most suitable pattern for the similar uses of the medium in Afghanistan.

Procedure

The basic hypothesis to be tested was that television has provided a demonstrably effective way of disseminating fundamental education and eradicating illiteracy in developing countries. To test the hypothesis, a historical review of the variant uses and effectiveness of the medium in selected developing countries, was made. And to establish a logical ground for recommendation, the elements of similarities and dissimilarities among the developing nations were explored. Finally the gathered information was related to present educational problems of Afghanistan.

Results

(1) The study disclosed a substantial body of empirical evidence in support of the hypothesis that television constitutes a viable instrument of education for meeting a wide variety of cultural and social needs in developing countries.

(2) In spite of their widely dispersed location over the world and the diversity of their history, culture, and political organizations, the countries surveyed exhibited certain striking similarities.

(3) The conditions in and the needs of Afghanistan were substantially the same as in the developing countries studied.

Recommendations

Regarding the use of educational television in Afghanistan, the investigator suggested that television can bring about a crucial change in education in Afghanistan; and that the medium seems to be a contributory answer to the problems that handicap education and consequently hold back the country.

Recommendations regarding the most suitable method of utilizing the medium in Afghanistan were made bearing upon the system of its operation, the subject matter to be broadcast, and ways of handling the initial problems. The system of broadcasting has been already decided as being governmentally owned and operated system. The subject matter should concern the areas where the most pressing needs in education are felt, namely, in elementary and adult education. Designating the elementary level (as distinguished from higher education) as the target area kills two birds with one stone in that subject matter of elementary level can be easily understood by adult illiterates as well.

With respect to the handling of initial problems, a start on a small scale was recommended, to be carried out as a joint responsibility of the Ministry of Education and the Ministry of Culture and Information. It was further recommended that Kabul, the capital of the country, should be considered as the first experimental ground for this project, not only because most of the developing countries started in this way, but because of geographical and demographic considerations, as well as, certain other significant considerations included in the text.

PREFACE

The investigator is indebted to Professor Charles A. Siepmann for his patience and sympathetic guidance generously given at all points during the development of this thesis. He has also benefited greatly from the advice of other members of his sponsoring committee, professor Frederick Redefer, and professor Emilio Guerra, as well as from the suggestions and cooperation of associated professor George Gordon.

For assistance in the collection of data, the investigator is under obligation to a great number of American organizations, such as Stanford University's Department of Communications Research, United States Office of Education, and the Peace Corps headquarters in Washington, D. C. for the magnificent help they have given to the investigator. He is also thankful and obligated to the individuals and organizations, outside the United States of America, that were prompt in responding to his correspondence, and helpful in providing valuable materials required for this dissertation. Among these organizations were Nippon Hoso Kyokai (NHK), Central Educational Television Overseas (CETO), and the United Nations Educational, Scientific and Cultural Organization (UNESCO) and its Department of Mass Communication.

Last, but not the least, mention should be made of the generous assistance of the staff of the Education Library, New York University, Mew York Public Library, Library of Teachers College, and Library of Ford Foundation. The laborious task of assembling the material upon which the thesis is based was greatly facilitated by their cooperation.

CONTENTS

CHAPTER ONE

INTRODUCTION

I. THE NATURE AND SCOPE OF THIS THESIS

The purpose of this study is to make an objective, judgmental evaluation with respect to desirable uses of educational television in Afghanistan, based on a comparative study of uses of the medium in developing countries. The specific problem, therefore, is to make a study of varied uses of educational television in selected developing countries through December 1965, and, on the basis of the facts gathered, to evaluate the appropriateness of similar uses of television in Afghanistan.

The project may be divided into three sub-problems as follows: (1) to give an account of the reasons for the use of educational television in developing countries, of the techniques utilized, and of such evidence as there is of the effectiveness achieved by the use of the medium; (2) to identify the similarities and dissimilarities (in terms of geographical, cultural, religious, and economic factors) between the selected developing nations, that have experienced educational television, and Afghanistan in order to evaluate the problems that may be involved in the similarities and differences between them; and (3) to relate the gathered information to the present educational problems of, and to the possible uses of educational television in, Afghanistan and thus to be able to draw a conclusion or make possible recommendations.

Delimitations of this study are as follows:

(1) This study will be limited to the utilization of educational television in developing countries up to December 1965.

(2) This study will be confined to those developing countries from which data, contributory to the purpose of this study, can be secured.

(3) This study will be limited to the use of television as an educational medium, thus excluding commercial television (other than as used as a vehicle for educational programs) and other educational media as well.

The basic hypothesis in this study is that television has provided a demonstrably effective way of disseminating fundamental education and eradicating illiteracy in developing countries.

The investigator has made two basic assumptions: firstly, he assumes that judgements cited in this study regarding some of the uses of television for educational purposes in developing countries, have been made with accuracy by qualified observers, and that their reports are reliable; secondly, he assumes that facilities for educational television on a national basis will be available in Afghanistan within the next decade, and will be employed for the purpose of fundamental education.

Certain terminologies have been used in this study and have been defined by the investigator to fit the concepts used throughout this thesis. They are as follows:

(1) Educational television refers to the uses of television "to cover almost any sort of educational video program pre-

sented for any serious purpose or in an attempt to teach something to someone."[1]

(2) Developing countries are those in which manpower is relatively unskilled, natural resources are almost unexploited, and literacy is confined to from two-thirds of the population down to a small minority.

(3) Literacy means sufficient amount of "reading, writing, counting, and learning many skills of practical value through communication media whose substance is derived from the experience of community, economic, technical and social development"[2] to help individuals in exploring ways of improving their living standards.

(4) Fundamental education has been defined by UNESCO as "that kind of minimum and general education which aims to help children and adults, who don't have the advantage of formal education, to understand the problems of their immediate environment and their rights and duties as citizens and individuals, and to participate more effectively in the economic and social progress of their community."[3] This definition indicates the variety of educational activities carried on throughout the world in such areas as mass education, social education, community development, and literacy campaigns.

---

1. George N. Gordon, Educational Television, (New York: The Center for Applied Research in Education Inc., 1965) p. 2.

2. M. Anas "Speech at World Congress of Ministers of Education in Teheran", Irfan (Kabul), No. 7 (August 1965), p. 83.

3. A. J. Foy Cross, Seminar on Visual Aids in Fundamental Education, Report of the U.S. participant in UNESCO (Messina, Sicily, August 31-September 26, 1953) p. 1.

II. RESEARCH PROCEDURES USED IN THIS THESIS

Since sub-problem 1--for what reasons, how, and with what effect has television been used for educational purposes in developing countries?--as well as sub-problem 2--what are the major similarities and differences, as related to the use of educational television between Afghanistan and the developing countries treated in sub-problem 1? are "an investigation of the functional uses of evidence and possibilities for applying its collected data to current issues and problems"[4], the historical research method is applied throughout both sub-problems in this project. Sub-problem 3 is a conclusion and recommendation based on the facts gathered and analyzed from sub-problem 1 and sub-problem 2. As the historical process requires, primary sources for sub-problem and sub-problem 2 were utilized as extensively as possible, and the secondary sources were consulted only in the absence of pertinent primary materials and to provide additional supplementary information. Because the available original source materials constitute a considerable portion of the study, the secondary sources have been used sparingly.

The information obtained from primary sources was reviewed and the relevant data were cross-validated. The authenticity of the data thus obtained was verified by personal interview and correspondence, as well as by other written and documented sources. Judgments, evaluations, and criticisms were carefully noted and examined for bias, contradictions, over-statement

---

4. C. V. Good and D. E. Scates, Methods of Research, (New York: Appleton-Century-Crofts Inc., 1954), p. 641.

and under-statement, and were appraised according to similar cross-validations. If independent verification of such statements was not available, caution has been exercised in reaching conclusions.

The collected data were surveyed and checked by the investigator to insure that all pertinent sources have been included and only reliable works have been employed for the solution of the problem. The data were scrutinized to make sure that they comprise sufficient material, analysis, and interpretation for the solution of the problem, with relevant material organized under the heading of each proposed area in a logical manner. The data have been subjected to recognized scholarly procedures of external and internal criticism as established by Van Dalen.[5]

After the process of gathering and cross-validation, the data were classified in accordance, with their reliability, and conclusions drawn from them were modified accordingly. The analysis and synthesis of the facts have been integrated in logical order in such a way as to present as full and complete a picture of the subject as possible.

For analysis, treatment and presentation, generally the following three major steps of historical research suggested by Good and Scates were taken into consideration throughout the thesis:

(1) Collection of data, with consideration of documents and remains or relics of primary and secondary sources of bibliographical procedure and of organization of material.

[5.] D. B. Van Dalen, Understanding Educational Research, (New York: McGraw Hill Book Co., Inc. 1962), pp. 260-261.

(2) Criticism of the data collected including the process of external criticism and internal criticism.

(3) The presentation of the facts in readable form involving problems of organization, composition, and interpretation.[6]

The interpreted and integrated facts are presented in chronologically ordered sequences to demonstrate the orderly transition of the use of educational television from the stage of an idea to the stage of practice and reality. Tables, diagrams, statistical charts, and any illustrative techniques that add to the clarity of the study were used when necessary. A special attempt has been made throughout this thesis to insure that the complete picture of the study should reflect as objective, prudential and logical a conclusion to the problem as possible.

III. SIGNIFICANCE OF THE STUDY

To indicate the significance of this study it seems appropriate briefly to survey the role of television in mitigating educational problems in developing countries, and to explore current educational problems in Afghanistan and the significance of possible use of educational television to meet those problems.

A. The role of television in solving educational problems in developing countries:

One of the most difficult problems facing two-thirds of the world's population--the so-called developing countries--is the problem of education. Knowing the importance of this problem, educational leaders in the developing nations are determined to

---

6. Good and Scates, loc. cit., pp. 179-180.

achieve mass education and provide themselves with at least a minimum level of literacy and fundamental education whereby they can improve and facilitate their present living conditions. They are aware of Whitehead's warning:

> When one considers in its length and in its breadth the importance of this question of education of a nation's young, the broken lives, the defeated hopes, the national failures, which results from the frivolous inertia with which it is treated it is difficult to restrain within oneself a savage rage. In the condition of modern life the role is absolute, the race which does not value trained intelligence is doomed. Not all your heroism, not all your social charm, not all your wit, not all your victories on land or at sea, can move back the finger of fate. Today we maintain ourselves and tomorrow science will have moved forward yet one more step and there will be appeal from the judgement which will then be pronounced on the uneducated.[7]

They are also aware of the relevance to them of Jefferson's famous saying that "if a nation expects to be ignorant and free, it expects what never was and never will be." They realize that the gap between their nations and more developed countries is wide and will grow wider unless they speed up their educational process by every means available to them. They believe that they cannot afford to delay this progress and thereby stifle their aspirations to transform their traditional pattern of life into a more

---

[7] Alfred North Whitehead, The Aims of Education, (New York: Macmillan Company, 1929), p. 25.

socially and economically advanced one. Their determination to achieve these goals is firm and strong.

Leaders in the developing countries realize that they have to run before learning how to walk and this process of growth has been, as far as education is concerned, facilitated by the new technology of television. The use of television in education has already partially fulfilled this aspiration in a number of developing countries. Television has already practically demonstrated its potentialities for solving educational problems, both formal and informal, effectively in many cultures around the world.

Television with its enormous capacity for reaching large audiences simultaneously and with great impact, has been notably efficient in improving the quality of various sorts of educational endeavours, as well as in reaching the masses regardless of number or physical distance with messages of significance regarded by its disseminators as educational. In addition to its use in instructing the masses it has proved itself to be instrumental in spreading a message instantaneously to nearly a total population and this is why over sixty of the so-called underdeveloped countries now have television systems and why the rest of them are likely to have it in a few years.[8]

B. Educational problems in Afghanistan and the significance of possible use of educational television to meet those problems:

Judging from the present growing educational problems in Afghanistan the outlook for solving them seems bleak unless

---

8. William P. Dizard, Television: A World View, (New York: Syracuse University Press, 1966), p. 15.

educational television is utilized. Public demands for more and better education in the face of a shortage of teachers, buildings, libraries, textbooks and other educational facilities (such as geographical maps and audiovisual aids) make the already existing pressure on unprepared Afghan educational institutions almost intolerable.

As a result of public outcry for more and better education in Afghanistan, increased enrollments have been such that it was virtually impossible for the experts of a combined committee of United Nations' Education, Scientific, and Cultural Organization (UNESCO) and Economic Commission for Asia and the Far East (ECAFE) to predict figures for enrollment sufficiently accurate to permit adequate future planning. This committee predicted in 1963 that the number of elementary school students in 1965 would be from 275,000 to 303,000. By 1964 this prediction had been exceeded considerably.[9]

Woeful inadequacies in the number of teachers, libraries, textbooks and even school buildings and educational facilities have resulted in serious problems which affect the fundamental quality of education in Afghanistan.[10] Educational and cultural authorities of the nation of course feel obligated to explore the most effective means to remedy these difficulties. Although such problems are not new in the experience of other similar developing countries, their intensity in Afghanistan appears to be as great or greater than in most others.

---

9. Aref Ghousi, "The Role of Education in Development of Economy", Irfan: Kabul No. 9, January 1966, p. 27.

10. United Nations Education Mission, A Report of the Mission to Afghanistan, (Paris: UNESCO, 1952), p. 20.

The Afghan authorities have already been in contact with some of the economically advanced countries to explore the possibilities of introducing television in the country. This study is therefore both important and necessary at this time, particularly in view of the country's educational needs and the many decisions regarding the practical employment of educational television which must shortly be made there.

The present study may also be of value in providing guidelines to administrators and broadcasters in other countries like Afghanistan which have not had previous experience in educational television or in telecasting of any kind, and which suffer from similarly deprived and under-developed elementary and secondary school facilities. The study should serve to bring out the extent of similarity or dissimilarity in the use of television in the various countries studied and thus provide a comprehensive picture of the extent and nature of its use from which other countries presently without television may profit. It is the first comprehensive study of its kind embracing most of the developing countries from which sufficient material can be secured.

CHAPTER TWO

TELEVISION

Scientific explorations have been enormously accelerated since World War II. Man's achievements in the diversified fields of science and technology since 1945 far exceed all of his efforts in this direction from the beginning of human history until then. This has been the case with electronic media of communications as well.

Less than a century has elapsed since James Clark Maxwell, a Scottish mathematician, predicted the action of electromagnetic waves. This was in 1873. Fourteen years later, Dr. Heinrich Rudolph Hertz was able to demonstrate not only the existence of such waves, but also that they can be sent through space with the speed of light. The experiments were continued by Sir Oliver Lodge and other scientists until Guglielmo Marconi, eight years later, in 1895 successfully sent and received wireless signals at home in Bologna, Italy. Marconi's West-East transatlantic wireless messages, sent from Glace Bay to England, made the idea of radio a reality. Radio thus was made publicly available in the first decade of the twentieth century.

Scientific explorations have not only been speeded up with the passage of time but have also become interrelated. These relationships in the various fields of science have become so complicated that virtually any discovery must be said to be a shared discovery rather than the results of individual efforts. Discoveries are now the product of many people working in different or related fields. The invention of modern television is an example of such interrelationship. For the invention of television cannot be credited to a single person, as Alexander Graham Bell and Guglielmo Marconi are respectively credited with the invention of telegraph and radio. It is rather a product of long evolution in the field of electro-communication devices preceding even the invention of telegraph. It is the result of the combination of a number of isolated discoveries in the field of electricity, electromagnetism, and electro-chemistry.

Although public use of the medium was inaugurated in London in 1936, the actual research on television goes back almost as far as the second half of the nineteenth century by scientists of different nationalities.

> The Alexandra Palace experiments were the culmination of half a century's research for a practical TV system by scientists of a half-dozen nations. They included the German, Paul Nipkow, who developed the first device for mechanical scanning of images in 1884; the Russian, Boris Rosing, designer of the first system of using a cathode ray tube as a receiver in 1907; and Vladimir Zworykin and Philo Farns-worth of the United States, who brought the potential of all-electronic television to working reality during the early 1930's.
>
> The United States was by far the most active area of experimentation; by 1932, the Federal Radio Commission had authorized twenty-five experimental television stations. Earlier the Bell Telephone Company had demonstrated wire-circuit television

by transmitting pictures from Washington to New York.[11]

In its early period of development television in the United States of America was considered primarily a commercial venture with a promising future in commercial communication where television might be used as a television-telephone to transmit pictures, charts, blue prints from one distantly located company to another. The outbreak of World War II halted, among other scientific activities, the growth of the television industry. Television development was resumed when peace was established. European countries were too wrapped up in their postwar political and economic problems to focus on television broadcasting; thus the postwar growth of television in Europe was slower than in the United States. As a matter of fact, due to the repercussions of the war very few European countries were able to resume broadcasting before 1950. British television, which had been initiated in 1936, was resumed in 1946, broadcasting to 1,300 sets, a figure which is very low as compared to its 20,000 prewar sets. France continued its telecasting in 1944, after Paris was liberated, and the USSR's postwar telecasting was resumed in 1946. By 1950, however, more than thirty European countries had already planned for television utilization, and within two years after 1950, the medium was introduced to developing countries of Latin America. Thus, in thirty years after its first public inauguration television has become a vital medium of information, education and entertainment throughout the world.

---

11. Dizard, op. cit., p. 23.

In 1950 there were no television receivers in the developing countries of Asia, Africa, and Latin America, whereas today more than sixty countries in those developing areas make use of television services. The world's 11,000,000 receivers and 130 transmitters of 1950 had risen to 130,000,000 receivers and 2,380 transmitters in 1963.[12] At the end of 1965 there were roughly 182,000,000 television sets in operation around the world.[13]

Although initially television, like its predecessor radio, was regarded as a state monopoly and therefore predestined to follow the traditional system of radio broadcasting, the strict concept of state monopoly control over television stations was moderated, as time progressed. The existing television stations of the world have, however, adopted one of three systems of operation: (1) state controlled non-commercial; (2) commercial; and (3) combination of both systems.

The development of television on a world-wide scale has passed rapidly through two stages: the first stage includes the medium's development in Europe, North America and Japan; the second stage encompasses the spread of television to the developing countries of Asia, Africa, and Latin America. Since most of the countries in the second stage of television's development have followed the patterns and systems set by the economically more advanced countries in the first stage, it would, therefore, be expedient to briefly survey the medium's development in some of

---

12. World Communication: Press, Radio, Television, Film, (Paris: UNESCO, 1964), p. 34.

13. For comprehensive detail about the increase of receiving sets and of operating stations in the world, refer to Appendix A.

the countries in the first stage, prior to surveying the growth of television in developing countries.[14]

I. THE MEDIUM'S FIRST STAGE OF DEVELOPMENT

A. UNITED STATES OF AMERICA

Although the United States had been more active, after World War II, in television than its European counter-parts, the medium's development would have been more rapid in the United States if it had not been slowed down by lack of precise government policy on technical standards, the fiscal obstacles of depression, and the broadcasting industry's concentration on its highly profitable radio operation.[15] In 1940, the Federal Communication Commission (FCC), approved a plan for commercial television broadcasting. Station operation in New York was begun in 1941 by National Broadcasting Company (NBC) and the Columbia Broadcasting System (CBS), on a fifteen-hour weekly schedule. By 1942 there were eight other stations of which six continued broadcasting throughout the war.[16] American postwar pioneers in broadcasting were NBC, CBS, and DuMont which entered as a third pioneer in 1944 but ceased operation in 1955.

It was not, however, before 1948 that television was fully accepted by the audience.

---

14. The world-wide use of television in education is discussed in Chapter III and Chapter IV.

15. "The Evolution of Television, 1927-34", Annual Reports of the Federal Radio Commission (FRC) and Federal Communication Commission (FCC), Journal of Broadcasting, Summer, I960, Vol. 4, No. 3, pp. 199-220.

16. Dizard, loc. cit.

> Comedian Milton Berle in 1948 became the first important star to have a regular program and his tremendous success lifted television from its previous status as a medium for amateurs to a position of professional entertainment. Television advertisement rates were increased from 25 percent to 50 percent in 1948.[17]

106 stations had begun broadcasting by September 1948, and the number would have increased if the FCC had not declared a freeze on the construction of new stations on the ground of technical problems regarding signal interference of stations broadcasting in the same area. Television receiver production increased from 178,000 in 1947 to 6,000,000 in 1952, and when the freeze was relaxed, the number of receivers jumped to 20,000,000 by the end of 1952.[18]

Another factor that marks the year 1948 is the discovery of color television. Work on color television had been started by CBS as early as 1940, undergoing the same process of growth as monochrome, until its discovery was announced by CBS in 1948. Shortly after this announcement, NBC declared its success in developing a pattern for color television. The struggle for the patent between the two companies was settled by the FCC which was acting as an arbitrator between the contestants. The patent was awarded to NBC on the ground that its system involved an electronic tube compatible with the reception of both black and white as well as color broadcasts.

---

17. "Television", Collier's Encyclopedia, XXII, (1965), p. 134.

18. Dizard, op. cit., p. 24.

By 1948, television had become the center of attention and had not only replaced radio as the most important means of home entertainment, but posed a serious threat to the movie industry as well. Thus the role of television as the new medium of mass communication was recognized in 1948. By 1962, there were 55,000,000 receivers in the United States. This number had increased to 70,000,000 at the end of 1965.

As a result of this phenomenal growth, television which was expected to be a medium for the upper and upper middle class in its early period, increasingly attracted a mass audience of middle, lower middle, and lower economic groups. By 1965 there were more than 540 commercial stations in the United States, privately owned and individually licensed by the FCC.

After the full establishment of a nation-wide television system in the United States, attention was focused on live global communication-channels through outer space. The Telstar I satellite that was launched by the National Space Agency on July 10, 1962, was the first practical and successful effort in the series of such attempts. The success of this endeavor made transatlantic communication, between the United States on the one side and England and France on the other, a functioning reality. Subsequently Relay I in December 1962, and Telstar II in May 1963 followed the operation of Telstar I.

When the people of Europe watched the September 15 Gemini II splash-down, they were watching one of the first successful uses of a portable

TV terminal live equipment mounted on the U.S. Guam.[19]

Undoubtedly the use of such equipment will some day allow television to be on the scene for an event practically anywhere in the world, and capture it live on portable equipment. The image could then be sent to a satellite from any given place or places in the world, and then relayed from the satellite through existing ground connections to local stations.

B. GREAT BRITAIN

In Great Britain, the British Broadcasting Corporation (BBC) had held a television monopoly until 1954, when the government approved the creation of the Independent Television Authority (ITA), a public corporation which went on the air in 1955 and whose programs are provided by privately financed program contractors who sell advertising space on the air. Compared to the commercial stations, BBC devotes a considerably higher amount of its telecasting to informational and educational programs. In 1959 there were ten million receivers in the United Kingdom, a figure then second only to the United States of America,[20] and this number was increased to 15,000,000 at the end of 1955.

As of 1963 there were sixty-six stations connected with the BBC as compared with twenty-two stations of the ITA network. The two networks with their affiliated stations have been successful in replacing radio and press as the most important mass medium of communication in Great Britain.[21]

---

19. "As You Like It", TV Guide, (New York: September 10, 1966), p. 2.

20. Henry H. Cassirer, Television Teaching Today, (Paris: UNESCO, 1960), p. 234.

21. Dizard, op. cit., p. 34.

C. FRANCE

French television went on the air in 1944 when Paris was liberated. In 1946 all political parties in France unanimously supported the legislation nationalizing radio and television in France. One reason for this support had been manipulation of prewar private radio and press by political and economic groups to the detriment of French national interest.[22] Thus, the system of government control was accepted and television became a part of Radiodiffusion Television Francaise directly under the Ministry of Information. This direct state control, which was part of France's traditional pattern, was modified in June 1964 when radio and television broadcasting services were transferred to an autonomous organization composed of an eighteen member board of directors representing government and people equally. This organization is called l'Office de Radiodiffusion Television Francaise (ORTF). ORTF assumed the responsibility of formulating the policy of broadcasting with the Ministry of Information serving in the capacity of advisor.

Television development has been comparatively slower in France than in other major European countries. In 1952, France had sixty thousand sets, the lowest ratio per family in any European country. In 1954, a plan for a network of forty-five transmitters was approved by the French National Assembly, though its execution was slowed down due to insufficient funds.[23]

---

22. Ibid., p. 27.

23. "Aspects of Television in Western Europe," Committee on Judiciary, House of Representatives, Washington, D. C., January, 1959.

In 1959, however, television transmitters covered roughly two-thirds of the country, broadcasting to approximately one million receivers. There were five million receivers in France by 1965, all non-commercially operated. The operating costs are covered by a license fee collected from the owners annually.

D. CANADA

Canada, like Britain, since 1954, combines commercial and non-commercial broadcasting in its system.

> Canadian radio and television broadcasting, like British, is a mixture of private and public operations. Again the process through which this system developed has been influenced by both technical and ideological considerations. On the technical side, the Canadians faced a problem similar to Soviet Russia, that of providing coverage over vast territory with relatively limited facilities. On the ideological side, the Canadian government was concerned with developing and maintaining a Canadian culture which would not be overridden by the United States (whose radio stations blanketed the Canadian border, where about 80 percent of the Canadian population lived) and which would provide broadcasts to suit the tastes of minority groups as well as the majority of Canadian listeners.[24]

When the government's restrictions were relaxed in 1959, an eight-station English language commercial network, CTV, was formed which started its broadcasting in 1961. CTV had eleven

---

24. Dizard, op. cit., p. 36.

stations in 1965, as compared with the sixty stations of the CBC. In spite of the fact that over half of CBC's radio and television network affiliates are privately owned and commercially operated, approximately seventy percent of its operating expenses are covered by annual government grants. CBC's affiliates are divided into two networks, one each for English and French language programs. Canadian broadcasting policy was always restricted regarding the use of the United States' programs on the ground that they are Americanizing the nation's culture. Following a policy originating in Great Britain, the Canadian government required that 55 percent of television program schedule be Canadian in origin. Despite this restriction, a survey which was conducted in 1965 showed that during prime evening viewing hours less than a third of the offerings were Canadian, and for the rest American programs were largely imported.1 In an attempt to remedy the situation and restrict American programs, the Fowler Committee was appointed in 1965 to study broadcasting. The Fowler Committee's report, recommending revision of the country's broadcast regulatory system, noted that the new system should have power to correct the serious imbalance resulting from trivial entertainment shows of the United States on the prime-time Canadian television.[25]

E. ITALY

Started in 1956, the Italian television which is controlled by Radiotelevision Italiana--a private joint-stock company owned largely by the government--covered the entire country and was serving a large portion of the Italian population through one mil-

---

25. Ibid.

lion receivers as early as 1959. The number of receivers had increased in 1965 to over six million sets, while the size of the audience had jumped to approximately thirty million persons. The reason is that most of the audience receive telecasting through television sets in the bars, coffee-houses, public squares, and other outdoor services. Radiotelevision Italiana (RAI) has two channels in operation, one for a popular audience and the other at a more intellectual level. Both channels arrange their programs to suit their audience; i.e., the channel for the popular audience broadcasts quiz shows, films, and women's and children's programs, while the channel for the intellectuals includes recitals, operas, dramatic productions and documentaries in their programs. RAI is financed by the charges set on each receiver. The set owners pay the equivalent of twenty dollars for the privilege of owning a receiving set. The Italian television system has been influenced largely by West German's broadcasting approach.

F. WEST GERMANY

Although Germany had started its experimental television broadcasting in 1935 and Hitler, knowing the persuasive capacity of the medium, was encouraging telecasting, prewar broadcasting in Germany remained limited to radio. The postwar television in West Germany developed on a regional self regulatory system based on the policy set by the Allies in 1945 emphasizing regional systems of broadcasting in all conquered nations. This idea was based on the desire to prevent centralized broadcasting in Germany such as that under Hitler and his party.

The first postwar television in West Germany started on Christmas Day, 1952 by Nordwest Deutscher Rundfunk (NWDR) station and in two years television stations were broadcasting all over the country. The general policy of each station on administration, fiscal affairs, and program operation is set up by a chief supervisory organ, a broadcasting council called Rundfunkrat or Hauptausschuss. The members of the council represent political parties, labor and business groups, cultural and educational, religious and professional communities. They select the station's administrative board that selects the station's manager. To secure their local independence and to facilitate handling relations between individual stations in network programs, the broadcasters in West Germany set up a "national consortium" called Arbeitsgemeinschaft der offentlich-rechtlichen Rundfunkanstalten der Bundesrepublik Deutschland (ARD), which by 1960 was established as the first television network in Germany. The stations were financed by monthly license fees charged on each television set, but since 1965 the fund has been supplemented by commercial advertising incomes. Up to 1963 ARD was the only network in the Federal Republic of Germany. However, the need for a second network to compete with ARD was felt and the German broadcasters organized a new network called ZDF-Zweites Deutches Fernsehn which went on the air in 1963.[26] There were more than eleven million receivers in West Germany at the end of 1965 and thus the medium is becoming the most accessible and most influential of all Germany's media of mass communication.

> Commercial pressures on programming are a negligible factor because TV advertising is completely separated from the rest of the program schedule; the commercials are all crammed into a single twenty-minute period each night except

[26] Dizard, op. cit., p. 40.

Sundays. Political pressures are blunted by the autonomy of each station and the check-and-balance factors built into its administrative structure. Decisions on the scope and content of German TV programming are largely in the hands of professionals. The result is a television schedule that provides, over and above the usual diet of sports and light entertainment, an unusually high percentage of programs designed to serve the upper level of its audience's tastes. At times this has involved controversial programs which were, politically and commercially "bad box office." West German TV has, for instance, taken the lead among all national media in dramatically depicting the Hitler regime and, in particular, the fate of German Jews under the Nazi rule. At another level, German television is a leading patron of modern drama, not only of Germany but from throughout the Western World. It is safe to say that the ARD network has presented more plays by the best contemporary United States playwrights in recent years than any American network.[27]

G. U.S.S.R.

The importance of television has been felt by the leaders of the Soviet Union and the government has decided that television is worth its heavy capital investment. In spite of the fact that providing relay facilities for a full linking of the television system over a vast and extensive territory has been a real problem, the suc-

27. Ibid., pp. 40-41.

cessful launching of the Soviet Union's communication satellite, Molniya 1, in April 1965 that transmitted live and color pictures from Moscow to Vladivostok, will undoubtedly solve this problem much earlier than their own deadline of 1980. Through its more than hundred key stations connecting approximately 290 large and small relay transmitters, broadcasting to almost thirteen million receivers, and serving more than forty million people daily, the Soviet television system has already become the most popular medium in the Union of Soviet Socialist Republics (USSR).[28]

Although, according to the USSR's historians, the first television image was transmitted by a Russian biologist, P. N. Bakhnetev as early as 1880, and the first television system using cathode ray tubes as receivers, was designed by Boris Rosing of the St. Petersburg Institute of Technology in 1907, experimental television transmission began in Moscow in 1939.[29] The medium's development, however, was suspended during the World War II and was resumed in 1946.

The postwar expansion of television was rather slower in the Soviet Union (perhaps because of either war repercussions or Stalin's bias in favor of printed media of communications).

> It was only after Stalin's death in 1953 and the liberalization of Soviet policies restricting consumer goods production that television really began to develop in the USSR as a mass medium. Even when the increased emphasis on consumer goods was

28. USSR in Figures, (Moscow: Central Statistical Publishing House, April, 1965), p. 102.

29. A Survey of Soviet Communication Electronics, Paul E. Green (ed.) (Cambridge, Mass: Lincoln Library Group Report, 34-76, M.I.T., (September 1958), pp. 16-18.

> cut back again later, the output of television sets continued to expand and their increasing availability became one of the few tangible evidences for Soviet citizens of the new liberality of the regime. In 1959 when prices of many luxury items were raised, the cost of television sets was actually lowered. The market is still by no means saturated, and even in the cities those who have sets of their own are in the minority. There are, however, enough sets in workers' clubs, hostels, and other public places to make television viewing a fairly common experience for the inhabitants of the larger cities... Far from being restricted to the transmission of Pravda editorials, Soviet television programming began to emphasize sports and other entertainment subjects during the Khrushchev period. Although normal precautionary steps were taken to see that the medium did not violate any ideological norms, there was little apparent effort to exercise close control of programming or to exploit television more assiduously as a propaganda weapon. There were no news programs on Soviet television until 1957, and then only once daily.[30]

The guidance of radio and television broadcasting in the USSR is the responsibility of a State Committee for Radio and Television Broadcasting, connected to the USSR Council of Ministers. The medium's responsibility, however, is dispersed among

---

30. Dizard, op. cit., pp. 181-83.

local stations, which have considerable autonomy in adapting their programs to regional needs.[31]

Television operation in the Soviet Union has been the target of criticism by Soviet officials as well as by foreign observers. The criticism, however, has been of poor programming rather than of technical aspects of Soviet television. Technically, the engineers of the USSR have made a series of important contributions to the development of the medium.

> Soviet engineers have, in fact made important technological contributions to television's development. The most dramatic of these was their design of the first system to transmit live pictures from an orbiting manned spacecraft. Their first communication satellite, "Molniya I" (Lightning), launched in April, 1965, transmitted color TV pictures from Moscow to Vladivostok shortly after it was sent into space. In 1964, Soviet engineers unveiled a color TV set that also provided three-dimensional pictures. The set displayed at the National Economic Exposition in Moscow, showed films of a railroad train that seemed to be coming out of the screen. Other engineers have experimented successfully with multiplex sound systems that permit television sets to receive audio signals in several languages--a potentially important development for multilingual nations like the Soviet Union and also

---

[31] Richard Tuber, "A Survey of Programming on the Central Studios of TV, Moscow, January-June, 1960", Journal of Broadcasting, (Fall, 1960), pp. 315-25.

for international transmissions from one language culture to another.[32]

Entertainment and informational programs are both included in the USSR's television programs. There are programs for women, children. Sports have been given special attention in the Soviet television system. Some of the impressive programs, on which the Soviet broadcasters pride themselves, are the telecasting of current plays, dance production, and motion pictures.

H. JAPAN

Moving from the North Atlantic area to the Pacific the great expansion in television industry has taken place in Japan. Like other highly developed countries, the Japanese had experimented with television before World War II. The postwar resumption of telecasting was late in Japan because the American occupation authorities rejected the petition of Matsutaro Shoriki and his group of businessmen, on the ground of Japanese economic problems. In 1950, the American authorities, however, agreed with the resumption of telecasting in Japan and formulated a new broadcasting policy that ended the traditional monopoly of broadcasting in Japan, allowing commercial television to compete against the government chartered Nippon Hosai Kyokai (NHK). Thus the Shoriki group group had the opportunity of starting their commercial television of Nippon TV Network (NTV) in 1952, when the American occupation had ended.

In the early period of development, the television industry's main problem in Japan was to increase the 3,000 sets of 1952 in a relatively short time. The problem was solved successfully, and

---

32. Dizard, op. cit., p. 188.

among the important factors that contributed to the solution of the problem were: (1) installation of TV receivers in public places to demonstrate the medium to the public on one side, and its significance to advertisers on the other, in order to persuade them to sponsor programs; (2) the telecasting of the wedding ceremony of the Japanese crown prince with a commoner's daughter, an episode in Japanese tradition that doubled the number of receivers; (3) the display of baseball and Sumo, Japanese sports of national interest.[33] In 1964 though the number of receivers was still only around 16,000, virtually millions of people had access to television through outdoor receivers.

The aforementioned factors stimulated enthusiasm to such an extent that in a relatively short time, television entered the home of millions of people throughout the country. In 1959 twenty six stations were broadcasting to 1,500,000 receivers. By 1960 128 stations were serving 7,000,000 sets. The Japanese television system with its 20,000,000 receivers serving an audience of 90,000,000 persons at the end of 1965 was first, after the United States, in the world. The majority of Japanese television stations are still operated by NHK, which in its procedures and structure bears close resemblance to the BBC.[34]

II. THE MEDIUM'S SECOND STAGE OF DEVELOPMENT

The following pages of this chapter will present some of the highlights of the second stage of television development--the stage in which the medium is introduced to developing countries

33. "International," Broadcasting, August 5, 1965, p. 76.

34. Dizard, op. cit., p. 44.

of Asia, Africa, and Latin America--as well as the practice and prospects of regional and intercontinental networks, which inevitably will help the further growth of the medium in developing areas of the world.

The reasons for the spread of television in developing countries of the world have been numerous, and among the prominent ones would be (1) consideration of the medium as a symbol of prestige; (2) the demand of European communities, living in the former colonial areas, for television; (3) the commercial opportunities offered by the medium to local businessmen; and (4) the promising prospects of television for the firms in economically more advanced countries "where television had stablized at a fairly high level and where commercial television companies faced the problem of slow-down in the rapid rate of expansion which had characterized their operation in the 1950's. The only area of expansion for their operation was overseas."[35.] An equally important reason might have been the realization of television's capabilities, on the part of the governments in developing countries, for upgrading mass information and public education.

In 1950 television was confined to the industrialized countries of Europe, North America, and Japan; the developing world of Africa, South America, Asia and Oceania had no television system of any kind. It is since then that television has begun to spread to developing countries. And the speed of this expansion has been so spectacular that presently almost every country in the world either has television or, with the exception of few African enclaves, is planning to have it in the near future. Television

---

[35.] Dizard, op. cit., pp. 49-51.

already is practiced in more than sixty of the eighty free countries of Asia, Africa, and Latin America; of the rest, fifteen can quite conceivably be expected to have utilized television by 1970.[36]

A. LATIN AMERICA

Among the developing regions of the world, Latin America was the first "testing ground" for television operation. In a short span of time (1950 and 1951), four Latin American countries, Mexico, Brazil, Cuba and Argentina started their television broadcasting and thus took the initial steps of introducing television to their countries.[37]

There were, however, problems of turning the medium into an economically successful business, which primarily seemed to be a high expectation in industrially less developed countries such as those in Latin America, where with the exception of a small minority of the upper class no one could afford foreign imported television receivers. This problem slowed down the speed of television development in its early period of introduction.

By 1965, however, television was a well established medium in larger cities of most Latin American countries, and the sets were locally produced in some part of the continent as well. Latin American television systems are privately owned and commercial operated, with the exception of Columbia and Chile where they are government owned and operated.

The expansion and maintenance of television broadcasting in Latin America have been mostly due to United States in-

---

36. Ibid., p. 48.

37. Ibid., p. 51.

vestments in those countries. This investment helped the growth of television to such an extent that by 1965 every Latin American country had television with the exception of Bolivia and Paraguay which had plans for it. In 1965, there were twenty-four stations for 1,215,000 receivers in Mexico; thirty-three stations for 2,500,000 receivers in Brazil; six stations for 550,000 sets in Venezuela; and eleven stations for 1,500,000 sets in Argentina. Television sets were locally produced in Argentina, Brazil, and Mexico by 1965.[38]

B. ASIA

Television development in the developing countries of Asia has been slower than in their Latin American counter-parts. The process has been even slower in the Middle East, and particularly in South Asia. This inertia has been the result of either political instability, as in Indochina, or reluctance on the part of the authorities, as in India and Pakistan, the two largest and most densely populated countries in that area. The pioneers of television in Asia, however, are the Philippines and Iran, which adopted television in their countries as early as the 1950's. Later on, Thailand, Malaysia, Indonesia, the People's Democratic Republic of China, Korea, the Republic of China, Iraq, Lebanon, Saudi-Arabia, Cyprus, Egypt, Damascus, and in 1965 Greece, Jordan, and Turkey had plans for television broadcasting by 1965.

C. AFRICA

In African developing countries, Nigeria was the first to inaugurate its television system in 1959 with local and British capital. In 1960, along with the independence of Nigeria, the federal government set up the Federal Television Service and contract-

---

38. Ibid., p. 55.

ed with the NBC network of the United States to provide management and programming for its key station in Lagos. By 1965, there were eight programming and relay stations in four major cities of Nigeria serving a total of 25,000 receivers. The Nigerian system of broadcasting adopted by Rhodesia in 1960, largely confined its broadcasting to the white minority of the country. In 1964, the white minority government nationalized the station as part of its local apartheid program.[39] Following the Nigerian establishment of television, more than a dozen of Africa's independent countries started to introduce television into their culture. Ghana, Ethiopia, Kenya, Sierra Leone, South Rhodesia, the Ivory Coast, Sudan, Mauritius, Nairobi, Uganda, among others, utilized television. The only African country resistant to the medium has been the government of South Africa, on the ground that television might be a serious political threat to its racial policies. Over thirty transmitters however, were operating in fourteen African countries by 1965, excluding the Arab countries of North Africa.[40]

D. NETWORK INTERCONNECTIONS

Another factor that might play a vital role in spreading television in developing countries of the world is the emerging of inter-state and intercontinental networks for exchanges of programs. The possibilities of establishing such networks have already been demonstrated by some economically advanced countries. Global networks for exchanging programs would be important to larger countries for reasons of prestige as well as of commerce, and would aid smaller countries which would need

---

39. Ibid., pp. 69-70.

40. Broadcasting, January 11, 1965, p. 62.

foreign programs since they could not afford the cost of nationally produced programs.[41] The idea of global networks for such purposes is not a novelty. It was started long ago and has been practiced in numerous parts of the world up to the present. The future seems to be promising.

The first attempt in the direction of creating a regional interstate network was made before World War II when the International Broadcasting Union (IBU) was established to encourage international cooperation in radio broadcasting. This organization, however, failed to develop due to "nationalistic uses to which radio was put particularly in the field of international short wave broadcast."[42] The first postwar attempt at such a network was made in 1946 at the Brussels conference, where IBU was replaced by the Organisation International de Radiodiffusion (OIR). The members of this organization consisted of the countries of Western and Eastern Europe, with the exception of Great Britain. Since most European countries had no television system prior to 1950, the new organization was concerned mostly with radio broadcasting. In 1948, however, when the organization began to consider the idea for exchange of television productions between national networks, the tension between Eastern and Western European countries had become so tense that it led to the split of the organization itself. The Western organization of Europe, entitled European Broadcasting Union (EBU) was thus formed in 1950. EBU took over OIR's administrative office in Geneva and its technical center at Brussels, while OIR was moved to Prague where it became an all out communist transmission belt coordi-

---

41. Dizard, op. cit., p. 82.

42. Ibid., p. 81.

nating radio and television exchange between socialist countries of Europe.[43]

Although formal exchange of television programs in the series of international programs between Great Britain and France started with the establishment of EBU in 1950, a formal bilateral exchange of programs began in 1952, when both sides overcame the difficulties that were involved in the differences of their television tubes.

The idea of an inter-European network, however, was the result (1) of an already successful exchange of programs between major European countries, and (2) the development of the medium in other European countries. One of the main factors helping the idea become reality was the coronation of Queen Elizabeth II in June, 1953, relayed by numerous television transmitters on the continent to millions of viewers. This event led to the establishment of an inter-European network in 1954. Ten years later, EBU had 28 members and 32 associates, including major American networks as well as the United States government through the Voice of America.

The technical and fiscal problems of EBU have been solved or assuaged by cooperation resulting from the tendency of European countries toward regional unity. The problems would have been and will be more challenging in developing countries which have no relatively small and densely populated geographical area, high technical competency, or cultural compatibility.

---

43. "Origin and First Steps of the EBU Program Committee," EBU Review, (Geneva) No. 85B, May, 1964.

EBU provides programs for its members, but without any obligation for acceptance. The contents of these programs have changed from an emphasis on sports in the early period, to emphasizing newsworthy events as determined by each individual member. The events of international interest, such as the death of the Pope, the marriage of kings, activities of NATO, problems of the Common Market, and the United Nations disarmament meetings in Geneva, are newsworthy events. Aside from newsworthiness, political factors have also been of special significance to EBU members. They tend to prefer programs that reflect favourable aspects of their countries.

By 1962, a daily exchange of visual news had become a well established feature of the Eurovision network, and in the same year, Eurovision's cooperation with the United States brought to Europe the first transatlantic communication relayed by satellite, the Telstar I. In 1964, 29 national television systems of Europe made almost 8000 pickups of Eurovision news transmission.[44] The growth of Eurovision has undoubtedly emphasized European unity without the use of propagandistic techniques. It has proven the importance of its existence to such an extent that the possibilities of the use of a European satellite have been discussed since 1963. Since EBU will have a great role to play in this matter, its past valuable experience in exchanging programs by satellites will certainly enrich the network's experience in this field.[45]

Eastern European countries, on the other hand, formed their own network which covers an extensive area from East Ber-

---

44. EBU Review, (Geneva) No. 91B, May, 1965, p. 51.

45. "Eurovision", George Hansen, Telefilm International (Los Angeles), January, 1960, p. 27.

lin to the Urals, and is perhaps to be expanded in the near future to Vladivostok, running a linear distance of more than nine thousand miles. The Intervision network was formed in 1960 by the state television organization of Czechoslovakia, Hungary, Easter Germany, and Poland with an organization similar to Eurovision.

The live relay of the Soviet Union's first launching of a man in space that took place between Moscow and London on April 14, 1961, was the first reflection of a tendency on the part of Intervision to contact Eurovision and Western Europe. This relay was followed, later on, by a special transmission between Moscow and Rome that consequently resulted in more exchanges between the two networks.

Although geographical, political and economic factors are serious handicaps in other parts of the world to the development of similar networks, such regional networks will become a reality in almost every continent "within the next decade".[46] A regional network will be in operation in Asia by 1970 under the leadership of a Japanese television organization, in an attempt to strengthen their political and economic power in the Far East. The Asian broadcasting conferences in 1957, 1958, and 1960 in Tokyo were among the practical attempts made to move in this direction. The Conference in 1962 that was held in Kuala Lumpur, established itself as the Asian Broadcasting Union, though its formal existence began in 1964, with headquarters in Tokyo.[47]

Initially, the ABU's membership included broadcasting systems of the big nations of the Far East, South Asia, and the

---

46. Dizard, op. cit., p. 96.

47. Ibid., p. 97.

Middle East. ABU's charter members are: Australian Broadcasting Commission, Broadcasting Corporation of China (Taiwan), All India Radio, Japan Broadcasting Corporation (NHK), Korean Broadcasting System, Radiodiffusion National Lao, Radio Malaysia, New Zealand Broadcasting Corporation, Radio Pakistan, Philippine Broadcasting Service, and the UAR Broadcasting Corporation (Cairo).[48]

Asiavision is a corporation founded by a group of Japanese equipment manufacturers for the purpose of exploring the possibilities of connecting Japanese television networks with Far Eastern television systems. This connection will be extended from Japan to Korea, Okinawa, and the Philippines crossing over the Asian mainland through Saigon to Laos, Cambodia, Thailand, Burma, Pakistan and India. And since direct regional broadcasting, due to technical problems, will not be presently feasible, Asiavision will develop gradually limiting itself primarily to the exchange of taped and filmed programs before establishing binational microwave connections. Such connections will be practically in use in the foreseeable future, under the auspices of the Asian Broadcasting Union.[49]

The idea for a network connecting the Arab television systems has been present almost from the time of the introduction of the medium itself to that area. This idea, however, has been hindered by the political differences between Arab countries. The establishment of an "Arab Middle Eastern Network" connecting television stations of Syria, Lebanon, Kuwait, Iraq, and Jordan,

---

48. Ibid., p. 304, No. 17.

49. Ibid., p. 98.

was announced by the American Broadcasting Company (ABC) in 1963.[50] ABC has been active in assisting Latin American networks as well.

Although the idea for a formation of a network in Latin America had existed earlier, it became a reality in 1964, when the Central American Network (CATVN) started its operation, with Nicaragua, Costa Rica, Guatemala, Panama, Honduras, and El Salvador as its members. The network was set up under the guidance of American Broadcasting Company as early as 1960.[51]

The next step beyond the regional networks is the establishment of an intercontinental network that would link all parts of the world. Obviously space communication satellites assure the idea of an intercontinental network; the primary concern at the present time, however, should be the full development of regional networks. If the growth of regional networks is encouraged, it will help solve some of the more serious political, economic and cultural problems that the intercontinental networks may face.

The first proposal in this connection was made by the United States at the special EBU meeting in New York in 1962. This proposal however, was resented by EBU members on the ground that it was presented without first consulting privately with other EBU associates.[52]

---

50. Variety, October 2, 1963.

51. Dizard, op. cit., p. 99.

52. Variety, October 10, 1962.

CHAPTER THREE

EDUCATIONAL TELEVISION IN TECHNOLOGICALLY ADVANCED COUNTRIES

I. THE NEED FOR EDUCATIONAL TELEVISION

Our era is marked by speed and rapid change. Today's problems are new and almost exclusively specific to this new technological age, and they require new approaches and new methods. Old methods are almost powerless to solve the new and newly emerging problems of our modern living, including education. Since the survival of a nation is largely dependent on the quality and quantity of education it provides, ever better and more education should be readily available to those who seek it.

Although the importance of education has been realized, none of the countries of the world has been able fully to satisfy the increasing educational demands of its population. "The universities in most parts of the world are feeling the pressure of increasing enrollments of students and the need to teach these students to a higher level of attainment in circumstances where knowledge is expanding at an ever increasing rate."[53]

Responding efficiently to these problems necessitates time, energy, and heavy financial investments; it requires constructing new buildings, supplying more educational facilities and providing qualified teachers. The gap between educational needs and educational efforts would have grown wider if new technology had not introduced a new and challenging education-

---

53. L. P. Greenhill, Learning Resources for Higher Education" (London) Medical and Biological Illustration, October, 1964, Vol. XIV, No. 4, p. 255.

al means for closing it, the use of television. For television, as many believe, has the capacity both of improving the quality of education and extending it to all those who seek it regardless of their numbers or geographical locations. It is one means of materializing the idea of democratizing education.

> Everyone has the right to education. Education shall be free, at least in the elementary and fundamental stages. Elementary education shall be compulsory. Technical and professional education shall be made generally available and higher education shall be equally accessible to all on the basis of merit.[54]

Educational television, as commonly used, is a confusing term covering two related but distinct types of service. The first may be described as that which contributes to cultural advancement. John Walker Powell mentions three characteristics that distinguish educational television from other forms of educational broadcasting: "(1) this is television that you watch on purpose, selecting each program for what it promises ..., this is television you have to pay attention to; (2) this is television that invites you to do something, learn something, think about something and expects you to participate; (3) this is television whose purpose is to turn off the set and do what has been suggested: read, paint, discuss,...."[55]

---

54. Universal Declaration of Human Rights, article 26.

55. John Walker Powell, Channels of Learning, (Washington D. C.: Public Affairs Press, 1962), p. 6.

The second type of service has reference to television's role in and contribution to formal education, whether in schools, colleges and universities or in adult education, and involves all of the considerations appropriate to high-level teaching in any classroom-development of immediate as well as long-range educational objectives, appropriate use of materials and techniques, planned sequences of concepts and events, and well organized lessons in keeping with accepted principles of learning. Thus instructional television is more specific in its meaning i. e. confined to the organized teaching-learning process and is part of the formal instructional program of an educational institution.

There are three principal ways in which, so far, instructional television is being used: (1) enrichment; (2) team teaching; and (3) direct teaching.

Enrichment refers to the use of television as a resource supplementary to the instruction given by the classroom teacher. It is usually used for lessons presented periodically which include valuable supplementary resource material on subjects related to the principal content of the course of study. Such programs enrich the educational experience of students in ways normally beyond the reach of the teacher and classroom resources. Examples are performances of great drama, the presentation of national and international experts, use of expensive scientific material and so forth.

Team teaching refers to the use of the medium when the televised instruction is supplemented by the classroom teacher's follow-up work with the students. In this approach television is used as a major teaching resource, but does not attempt to do the total teaching. The expository phase of teaching (subject matter, principles and essential ideas) is presented by the studio teacher and the classroom teacher thereafter conducts activities to reinforce or clarify and extend the material to bring about maximum learning. The classroom teacher is an integral part of such use of television and only close coordination and cooperation between all the participants make the teaching-learning process a success.

Direct teaching involves the situation in which the studio teacher presents the total instruction. This has been done usually when no teachers are locally available for specialized subjects such as foreign languages, advanced mathematics and science, or when the number of qualified teachers is too limited compared to the large number of students needing the course. Well prepared study guides, diversified and extensive reading lists, and regularly scheduled written assignments or reference projects mailed to the school tend to encourage and stimulate the self-directed student.

Systems of transmission vary. It is usually the aims of the educators and the needs and circumstances of the schools that determine and condition the system of televised transmission they choose. Closed circuit and open circuit educational television are the two systems of utilizing television for transmitting educational materials.

Closed-circuit television is the electronic transmission and reception of images within a prescribed area, when the origination point of transmission and the receiving areas are linked by coaxial cables. Multiple, simultaneous transmissions are here possible, static and like interferences are excluded, as also is the general audience that has access to open circuit broadcasts.

Open-circuit, on the other hand, refers to broadcast situations where programs are radiated on the air for reception on any viewer's receiving set within the transmitting range of the station.

II. INSTRUMENTALITY AND SERVICEABILITY OF EDUCATIONAL TELEVISION

Prior to the use of educational television the importance and practicability of radio, film and other audiovisual aids in education had been known to many cultures. This situation helped utilization of television for educational purposes to appear a natural rather than a totally strange phenomenon. Educational television, though in many cases different from previous educational aids, would, as some believe, be more constructive if it were used in combination with facilities already in use such as the printed word, radio, film, and audiovisual aids rather than displacing them.[56] If it is used properly, television can certainly make a contribution to the dissemination of knowledge throughout the world unsurpassed by any other medium. Its great efficiency will be revealed more conspicuously if it is used by skilled authorities, for "tools other than in the hands of craftsmen are useless" and television is not an exception.

> There can be no question that, in this matter of educational telecasting, there are those who have rushed in where angels fear to tread. Television is not an unmixed blessing. It is a two-edged sword. Every technological advance does not automatically spell progress. Without the exercise of

---

56. Henry Cassirer, Television Teaching Today, (Paris: UNESCO, 1960), p. 161.

great intelligence and true conscience it may indeed prove a curse.[57]

There are certain distinctive advantages inherent in the medium. The first and most important one is television's capacity to move information in quantities and over distances and with an immediacy paralleled only by radio. This capacity in television is so enormous that it far exceeds the human capacity of both teachers and students to absorb it fully.

The second significant advantage, corollary to the above, is television's capacity to embrace audiences un-precedented in size, as unaffected by distance. Thus to localize the use of television is, on one side, to defy the natural genius of the medium as it functions to unify humanity.

In addition to the aforementioned advantages, television (in formal as well as informal education) makes access possible for all students to the best and most brilliant teachers wherever they may reside or be. Moreover, it encourages even the best teachers to prepare thoroughly, organize well, and present carefully and effectively their lectures, lecture-demonstrations and discussions, as it also helps teachers to illustrate their courses by making use of more graphics, films and other visual resources; it provides every student with a closeup view of the instructor; it puts every student in a front seat where the instructor gives the impression of lecturing directly to him rather than to a group.

One of the limitations of television, on the other hand, is the problem of one-way communication which is different from

---

57. Charles A. Siepmann, Television and Our School Crisis, (New York: Dodd Mead and Company, 1958), p. 115.

the normal process of human communication which is cyclical and includes reciprocal exchanges. This limitation, however, is shared by previous means of education such as books or lectures in large classrooms where there is little chance for questioning the instructor. This disadvantage can to some extent, and in certain situations be remedied by facilities for two-way communication. Similarly, technical limitations of the medium such as its lack of stereophonic sound, its monochromatic system, its confinement to a small field of vision, have either already been remedied or will be remedied when future adaptations, innovations and inventions are made.

> In supporting educational television, we need not be apologetic because it imposes some limitations on ideal teacher-student interactions. This ideal is rarely achieved by conventional teaching methods. What we need to do and what we can do is to define the problems and solve them where educational values justify the effort. The limitations on the interacting processes in teaching and learning by television can be solved by making adaptations within the full potentials and limitations of this medium, or by utilizing supplementary and auxiliary instructional procedure for which models already exist.[58]

---

58. John C. Adams, C. R. Carpenter, Dorothy R. Smith, College Teaching by Television; Report of a conference sponsored jointly by the Committee on Television of American Council on Education and the Pennsylvania State University at University Park, Pennsylvania, October 20-23, 1957 (Washington D. C.: American Council on Education, 1958) p. 10.

As to its effectiveness sufficient evidence has been accumulated to prove that educational television is at least as effective as conventional classroom teaching. More than 400 experiments in the United States alone have been conducted to compare educational televised studies with traditional classroom teaching. Most of these experiments showed no significant difference between the achievements of matched student groups taught conventionally and by television, and where there was such difference it was far more often a difference favoring the televised classes. Wilbur Schramm has reviewed 393 experimental comparisons of instructional television with classroom teaching which reveals that in 65 percent of the comparisons there was no difference in the amount of learning; in 14 percent the classroom students learned more as compared with 21 percent who learned more by television.[59]

As in all education, the need for further and more refined research in instructional television is eminently desirable. But the body of available evidence is sufficient to have made a prima facie case (supported, incidentally, by the informed opinion of the great majority of teachers who have effectively incorporated television in their teaching) for television's use as a viable instrument as aid in formal education. The above endorsement by participants in television's use in formal education is by no means confined to the United States--is indeed found in all the western countries that make use of instructional television. Significant for the purpose of this study is the fact that such expansion of

---

59. Wilbur Schramm, "What We Know About Learning from Instructional Television, Educational Television: The Next Ten Years, Stanford, California: Institute for Communication Research, (California: Stanford University Press, 1962) pp.52-76.

television's use in formal instruction has for the most part relied on such empirical evidence without awaiting the corroboration of more scientific measures of television's viability.

III. EDUCATIONAL TELEVISION IN INDUSTRIALLY DEVELOPED COUNTRIES

Uses of television in both formal and informal (cultural) education vary with the cultural tradition and conventional systems of organization, and more particularly with the systems of broadcasting and of education that obtain in different countries. As regards informal (cultural) education, this forms (though in varying degree) a staple component of television's program service in all countries where television is centrally operated, (whether under government or independent auspices) as a public service. It is only in countries (e.g. the United States and some South American countries) where commercialism and the profit motive have been allowed to dominate, that informal (cultural) education tends to constitute a minor or insignificant component of output. The countervailing force to this tendency in the evolution of television in the United States has been the development of local stations for non-profit, educational use on a reserved band of frequencies assigned by the Federal Communications Commission. But it is in formal (instructional) television that the variables between national systems are seen to best advantage as affecting the nature and the flow of instructional material, as illustrated below.

A. THE UNITED STATES

Initially, educators' access to the television airways was limited to channels occupied by commercial broadcasters. Many such stations made generous provision for school broadcasts (and in some instances for college courses) at times not already preempted for sponsored programs. But difficulties arose as, when a sponsor became available, time granted for instructional television was shifted to hours less convenient to the school systems involved, precluding that constancy and continuity in time on which scheduling of school broadcast hours essentially depended. It was only in 1952, when reserved frequencies for non-profit educational broadcasting became available, that instructional television came into its own. On these reserved frequencies instructional television secured an assured place at convenient times and was thus enabled to plan ahead and integrate school broadcasts with school curricula.

But, given the novelty, to say nothing of the costs involved in television's use, progress would have been much slower but for the initiative of the Ford Foundation--launching and financially supporting experiments designed to test television's contribution to formal education. The extent of such grants in aid ran into the tens of millions of dollars. Subsequent federal grants further assisted the growth of television's use in schools. It is estimated that 49 percent of schools in the United States currently make some use of instructional television.[60, 61]

Given the decentralized character of education in the United States, use of instructional television was initiated locally and remains largely such to this day. An exception to this rule was

---

60. "The Needs of Education for Television Channel Allocations", U.S. Department of Health, Education and Welfare, 0E-34017.

61. "Learning by Television", Fund for the Advancement of Education, 1966, p. 38.

the introduction of regional services as school lessons, given by picked teachers associated with skilled television production, were disseminated over six states by means of transmission from an airplane circling at 30,000 feet. The ultimate evolution of television's use in schools, as some see it, will involve centralized production of lessons of the highest excellence, combined with the highest standards of production skills, disseminated whether on video tape or by open or closed circuit broadcasting for reception in local schools at their discretion much as textbooks are chosen and distributed today.

Comparable, though less extensive use of instructional television has been made, at an experimental level, in higher education where research indicates that achievement by students receiving instruction by television is in most instances equal to that of matched student groups taught conventionally. As related to the ever rising enrollment of students and the difficulty of matching this with a comparable increase of competent faculty and of physical accommodation, use of television is seen as realising the following potential advantages:

(1) Students (as in San Francisco State College[62]) receive part of their instruction by television in their homes, thus alleviating classroom congestion.

(2) Required courses, involving the subdivision of students into sections taught by numerous instructors, can now be given by one instructor (and the best at that), thus relieving his fellow instructors for redeployment to such activities as conducting dis-

---

62. "Instructional Television Research", San Francisco State College, Dolores Press, 1958.

cussions in groups of manageable size, individual counseling, etc., etc.

(3) Higher education is enabled to reach would-be students whose circumstances preclude their regular on-campus attendance. Such students include housewives, the home-bound and (as in at least two instances) inmates of prisons! Thus over eight years the Chicago Junior College system has incorporated televised instruction with striking results in terms of the extension of participating students without involving hopeless congestion of its classroom and on-campus facilities--and this without sacrifice of academic standards attained. On the air 25 hours weekly, the "Chicago TV College" has served over 80,000 individuals in over 120,000 course registrations for credit. Non-credit enrollment has been consistently as high as that for credit. An "unseen audience" of over 200,000 viewers per course had already been reached by 1964.[63]

(4) Inter-institutional cooperation has resulted in the pooling of faculty on different campuses and the exposure of students on each and all of these to the finest lecturers in each.[64]

(5) In university extension work, where the content of course permitted it, itinerant lecturers have, by broadcasting on television, been saved thousands of miles of travel a year. Such saving of cost applies only to instruction in which 2-way commu-

63. Chicago Public Schools, "8 Years of TV College: A Fourth Report", a report by the Chicago Public Schools, September, 1964.

64. The earliest experiment of this kind was in Oregon and is reported on in "Inter-Institutional Teaching by Television, 1957-1964, Glenn Starlan and John Lukas: Oregon State System of Higher Education, Oregon, 1964.

nication with students is not essential to the educational goals attempted.

(6) Effective use of television has been made in the further in-service training of teachers by scholars and experts not normally available to the individual teacher training institutions.

(7) Through television talented High School seniors have been enabled to take college level courses and thus (subject to satisfactory attainment in examinations given) to secure advanced standing when matriculating for their college careers.

Unlike European countries, the United States' use of instructional television encompasses thousands of school systems at all grade levels and thus provides the most extensive precedent for consideration by developing countries as related to their needs.

> Beyond its value for American education, our ETV is a store-house of information for educators from Asia, Africa, and Latin America. Our experience cannot be exported wholesale, but (to cite only one example) the pacesetting South Carolina state education department's use of television to reach deprived children in rural areas offers useful lessons for any underdeveloped country.
>
> The United States has already taken the lead in exporting techniques abroad. For over a decade, American ETV stations have been a Mecca for foreign educators interested in the subject. Since 1960, American pilot program assistance to ETV efforts in underdeveloped countries has been

increasing rapidly. The Ford Foundation took an early lead with projects in India and in Latin America. The only training school for TV educators in the developing world is being operated under American sponsorship in the Philippines. More significantly, since 1963, the U.S. government's technical assistance program has included funds for ETV projects in developing countries. These efforts are still modest in scope, especially when measured against the problems with which they are intended to deal. Their usefulness lies in setting a new pace for educational development in the poorer countries.[65]

B. GREAT BRITAIN

In Britain education is decentralized, broadcasting centralized. Broadcasts to schools over radio date back to the first years of operation of the British Broadcasting Corporation which came into being in 1927. So successful were they that by the end of World War II, 73 percent of British schools were availing themselves of the BBC radio transmissions.[66] Radio's success in service to the schools is indeed regarded by some as one of the factors accounting for the belated and cautious introduction of televised school broadcasts. These were first introduced in Britain in a series of experimental telecasts transmitted by closed circuit to six secondary schools in the London area in 1952. It was not, however, until 1959 that the BBC announced plans for the first nationwide, permanent service of school television. Sim-

---

65. Dizard, op. cit., p. 19.

66. F. N. Lloyd Williams, "School Broadcasting, Sound", EBU Review (Geneva), No. 70B, (November 1961), p.44.

ilar services were concurrently offered to the schools through Britain's postwar alternative commercial television service, the Independent Television Authority. Both services have provided enrichment programs, neither of them adventuring into either direct or team teaching as developed in the United States. But the chief contrast between the use of instructional television in these two countries lies in the central production and nationwide transmission of these services in Britain--in contrast to the predominantly local organization and transmission in the United States.

C. FRANCE

In France both education and broadcasting are centrally organized and under government direction. Such a combination is perhaps ideal as it facilitates the fullest exploitation of television's unique capacity for the universal transmission of excellence. Centralized production of school telecasts allows of the use of the nation's best teaching and production talent. Centralized organization of the schools facilitates transmission to the largest number of schools with the least difficulty regarding scheduling.

French television programs for schools are prepared by the National Pedagogical Institute established by the Ministry of Education. They are produced by Radio Diffusion Television Francaise in consultation with educators at all stages of production. School transmissions were initiated soon after the postwar resumption of telecasting in France. By 1959 approximately 3,000 schools were equipped with television receivers. By 1961 program service was extended to 11 courses weekly serving 5,000 schools.[67] Prior to 1960 the service consisted of enrichment programs, but a wide variety of programs providing direct instruction were introduced thereafter. French developments of instructional television have greatly influenced other European countries, particularly those (like Belgium and the Netherlands) with state controlled television systems.

D. ITALY

Television in Italy is centrally controlled by Radio Televisione Italiana. School television programs were first introduced in 1954. Italy's proudest and best known achievement has been the organization of Telescuola--initiated on the 25th of November, 1958. This constituted the first attempt in Europe to use television for the direct instruction of mass audiences. In part the aim of this service was a frontal attack on adult illiteracy as on the lack of school facilities in rural areas. By 1961 Telelescuola was broadcasting 34 hours a week covering a wide range of subject matter. Viewers were provided with special textbooks published by the Ministry of Education. Some 2,000 listening posts were organized. The audience consisted predominantly of students, 1/3 of them high school dropouts.[68]

E. U.S.S.R.

Both broadcasting and education in the Union of Soviet Socialist Republics are under centralized government administration. Already in 1960 television had become one of the country's

---

67. Henri Dieuzeide, "10 Years of School Television", EBU Review (Geneva), No. 69B, (September 1961).

68. For the Early Development of this Service see "Telescuola Enters Its Second Year" by Maria Grazia Puglisi, EBU Review, No. 59B, January, 1960; also "Educational TV", by Italo Nervi, EBU Review, No. 69B, September, 1961.

most important media for the dissemination of technical and scientific knowledge. Its expanding service has paid increasing attention to educational television both for children and adults. Experiments in direct teaching proceed hand in hand with use of the medium for enrichment purposes.[69]

> In one 1962 experiment, fifty-two thousand farmers in the Moscow region were formed into listening groups to view a televised series of lectures and demonstrations on scientific agronomy. Soviet television authorities have indicated they plan to set up separate educational channels in major cities throughout the country; the first of these was inaugurated in Moscow in October, 1964. They have also tried a new approach toward ETV programs for children. Instead of broadcasting into classrooms enrichment lessons are transmitted to home receivers in the afternoon after school hours. The lessons are coordinated with the school curriculum in ways that assist students to prepare their homework and review the day's school lessons.[70]

F. JAPAN

At the second international conference on school broadcasting in Tokyo in April, 1964, broadcasters and educators were convinced of the fact that Japan is the first country in the world that has fully integrated television into its educational structure from kindergarten to university-level studies, as well as into the

---

69. Cassirer, op. cit., p. 229.

70. Dizard, op. cit., pp. 222-223.

broad field of adult instruction. The Japanese traditional emphasis on learning plus their tremendous advances in mass education in the past fifteen years have resulted in development of Japanese educational television to a greater extent than in any other country in the world. There are as many educational television stations in Japan as there are regular stations in any single western European country.[71] Both broadcasting and education in Japan are centralized.

School broadcasting by radio had been expanded in Japan to such an extent that by 1959, 99 percent of all schools were equipped with receivers. Thus the subsequent acceptance of television as an educational medium was a quite natural development. Japan's first experimental series of school television programs was started by their public service broadcasting system (NHK) in the summer of 1953. In 1958, NHK inaugurated a separate national educational television network. In 1959, approximately 10,000 Japanese schools were regularly receiving educational programs.

Televised educational broadcasting is the main involvement of NHK, though NTV--a commercial network--and the Foundation for the Promotion of Science and Technology are also actively participating in Japanese educational broadcasting. NHK operates two television networks, one for general cultural services and the other for educational programs. The former offers

---

71. Ibid., pp.42-43.

news, dramas, movies, informative programs such as programs for women and children, transmitting 18 hours of daily programs that are simultaneously broadcast over 168 stations throughout the country. Its services cover 87 percent of the country. The latter network, on the other hand, is exclusively for educational purposes, serving the following "four pillars": schools, education by correspondence, adult education, and cultural activities. Its daily programs of 13 1/2 hours are beamed from 156 stations, covering 86 percent of the country.

In 1964, NHK had 100 classroom programs for a total of 34 hours 20 minutes a week in television that were relayed by means of microwave facilities to its 156 affiliated stations, and used concurrently in classrooms throughout the country.

Nippon Educational Television (NET) which is operated by NTV produces school programs that are broadcast over 19 commercial stations with 72 transmitting stations that cover 80 percent of the country's population. It produces 41 school broadcasting programs weekly totalling 12 hours 40 minutes. Since school broadcasting policy is determined by the government the contents of the programs of both NHK and NTV are fundamentally different from each other.[72]

By 1965, there were 46 educational television stations transmitting programs throughout Japan largely operated by NHK and, on a smaller scale, by commercial companies as well. The NHK stations retransmit their educational programs over 238 smaller re-

---

72. Yoshikazu Kasuga, General Managing Director and General Director of Broadcasting, Nippon Hoso Kyokai (NHK), "School Broadcasting by Television in Japan", Proceedings of the Second International Conference of Broadcasting Organizations on Sound and Television School Broadcasting", op. cit., pp. 62-71.

lay satellite stations designed to reach school and home receivers in isolated areas. Almost all Japanese schools are now equipped with television receivers. Adult educational telecasting services are beamed to 18 million home receivers reaching an audience of over 90 million people. To advise NHK and its affiliated stations on educational programs, the teachers and school administrators have formed a National Federation of Study Group of Radio and Television. Prior to 1960 NHK was focusing its effort on enrichment programs, but since then direct teaching has also been emphasized.

NHK Radio and Television Culture Research Institute conducted a nationwide survey of the utilization of school programs between October 1 and November 22, 1965. The results of the survey are given below.

	Total No. of Schools		Ratio Schools to Receiving Facility	Schools With Receiving Facility	Ratio of Broadcast Use	Schools Utilizing Broad-casts
			%		%	
Kinder-gartens	7,996	Radio	90.5	7,240	38.5	2,790
		TV	86.4	6,910	76.5	5,290
Nursery Schools	10,910	Radio	76.9	8,390	30.6	2,580
		TV	82.0	8,950	83.7	7,480
Primary Schools	26,209	Radio	96.7	25,340	51.3	12,970
		TV	97.1	25,450	78.4	19,950
Secondary Schools	12,316	Radio	96.5	11,890	33.0	3,920
		TV	96.0	11,820	24.5	2,890
Daytime High Schools	3,937	Radio	96.6	3,800	32.0	1,220
		TV	90.2	3,550	9.2	330
Evening High Schools	2,313	Radio	88.9	2,060	18.5	380
		TV	75.6	1,750	5.6	100
Total	63,681	Radio	92.2	58,720		
		TV	91.7	58,430		

([73*])

[73(*)] Source: "Around the World", CETO News, No. 12, (September, 1966), p. 69.

CHAPTER FOUR

EDUCATIONAL TELEVISION IN DEVELOPING COUNTRIES

I. INTRODUCTION

Although the UNESCO's criterion is that "a country is inadequately supplied with information if for every 100 people there were fewer than 10 copies of a daily newspaper, five radio receivers, two cinema seats, and two television receivers", surveys of UNESCO experts indicate that almost all the developing countries of Asia, Africa, and Latin America have not been sufficiently provided with informational media to meet the above criterion. According to one of these surveys, "nearly 70 percent of the world's people living in over 100 countries in Africa, Asia, and Latin America, lacked the barest means of being informed of events in their own lands let alone in others."[74]

Television with its capacity to combine aural and visual elements, constitutes a complete medium intelligible to intellectuals and to illiterates alike. It is perhaps the most powerful and effective medium for mass education among all mass media of communication. Its persuasive potency in impregnating traditional patterns of living with new values and modes of thinking can, hopefully, be used as a revolutionary factor in changing attitudes of tradition-bound cultures.

---

[74]. UNESCO's Meeting of Experts on Development of Information Media in Latin America, Developing Information Media in Latin America, Santiago, Chile, February 1-13, 1961, Report of the Meeting, (Paris, May 31, 1961), p.5.

Although strict principles of economy might not seem compatible with the use of television in developing countries, particularly at their present stage of development, television has been introduced in over sixty such countries and "the rest of them will have it in a few years." Television's potential in improving educational and informational standards and its significance in revolutionizing the pace of development cannot be ignored.

> The proper use of radical new educational techniques--television included--is a crucial need for emerging countries. In developed societies, schooling takes on the aspect of a consumer good; we do not have to relate the study of Shakespeare's plays to the growth of national output. The poorer nations are obligated to regard their educational effort as a major capital investment. For the next generation at least, the problem is that of determining how to do this quickly and efficiently. Seen in this light, educational television may be the social and economic bargain of the century for the new nations of Asia and Africa. Despite expensive installations, it may be the cheapest per-capita-cost way of using "x" resource of teachers to get "y" amount of facts and "z" degree of twentieth-century attitudes into the minds of half the world's population now living in ignorance and poverty.
>
> The most obvious need is for childhood education. Under present development plans a few underdeveloped countries will achieve universal education during the next generation. Most of the

rest will require two generations or more. The bottlenecks in teacher training, school construction, and equipment are compounded in many countries by ill-advised reliance on traditional Western educational methods. However well such methods have worked in Europe and America, they need to be profoundly modified for the newer nations. The need is for effective educational shortcuts to eliminate a massive inventory of social and technological illiteracy. Shortcuts exist, and they are mainly the product of Western technology. They involve the whole range of audio-visual techniques, television included, which have given new pace and dimension to modern education in the West. There is, in fact, little hope that the educational needs of the new nations can be met without a generous application of these techniques.

Beyond the primary education of children, the re-education of entire societies must be considered. This includes "adult education" as we use the phrase to mean supplementing earlier education. It also means providing the masses of Asia, Africa and Latin America with a comprehensible image of their new societies. For most of them, the idea that they are citizens of a new country is largely irrelevant to their day-to-day lives. Their social loyalties are to a caste, clan, village, or religion. Most of the shining development plans drawn up in the capital city run against this stubborn fact. The older Western symbols of nationalism--flags, anthems, parlia-

> ments, and the like--are meaningless or inadequate to the task of establishing national identity. What is needed is what television and its audio-visual allies can best provide, namely a dramatic visualization of nation building, brought directly to every corner of the country in terms that can be understood by everyone. Television can show its audience the rest of the world beyond their borders, relating it to their own lives. It can tell its audiences that they have a stake in the new order, and demonstrate to them that the best of their older traditions can be preserved and enhanced through change. The medium's potential as a nation-building instrument has only begun to be explored, but it may well prove to have revolutionary implications for all the developing world.[75]

To the developing countries of Asia, Africa, and Latin America that are going through an active period marked by an effort to revolutionize the economic, social, and cultural lives of their populations, television, if wisely used, is the timely answer for the accomplishment of those aims. Far from being a luxury confined to technologically developed countries that can afford it, television may be so profitable an investment that its use will directly influence the economic and social growth of economically less developed nations.

The medium, as a matter of fact, is of a special value to developing nations where manpower is a prerequisite for nation-

---

75. Dizard, op. cit., pp. 16-17.

al development; where an enormous body of students have to receive instruction in the shortest possible time and in a limited number of institutions by a handful of qualified teachers and with meager educational resources; and when it has to be adapted not only to intramural teaching in schools and in colleges, but to teaching in widely dispersed groups and communities in suburbs of rural areas. The new innovations and developments in television such as Telstar and earth satellites that have made continental relaying programs a functioning reality are certainly measures of great hope whereby the flow of information and education will be facilitated. Whether it is used in formal or informal education, television can bring a significant new dimension to the modernization of the developing countries. It is potent in educating the mass in the fundamental elements of living as well as in the direct teaching of all subjects from kindergarten to university level.

Although television was introduced in some of the developing nations in the 1950's, active steps in utilizing the medium for teaching purposes have been taken only since 1963 when educators of the developing world had attended the two international conferences on school broadcasting sponsored by Educational Broadcasting Organization in Rome in 1961 and in Tokyo in 1964.[76] Most of the delegates from the developing countries after observing examples of the use of educational broadcasting at these two conferences became enthusiastic in applying these models to their own educational patterns. Their initial efforts, however, were not totally successful due to underestimating the problems of installing and maintaining television systems that were different in developed and developing countries. Neverthe-

---

76. Ibid., p. 232.

less, some significant results of utilizing television in educational activities followed.

II. EDUCATIONAL TELEVISION IN LATIN AMERICA

Some of the Latin American countries were pioneers in experimenting with educational television in the late 1950's. It was not, however, until 1961 that educators and broadcasters showed sufficient interest to set up a regional commission on educational television problems.[77] Most of the seventeen countries attending UNESCO's regional seminar on educational television in Mexico City in 1964 reported that plans regarding educational television were on their way.

A. COLOMBIA

Colombia is a Spanish speaking republic that is neighbored by Venezuela and Brazil on the east, Ecuador and Peru on the south. It extends up the Isthmus of Panama to the Republic of Panama with a coastline of 913 miles on the Pacific Ocean and 1,094 miles on the Caribbean Sea. The country's population, according to the United Nations estimate in 1963, is 15,098,000. Bogota is the capital of the Republic of Colombia. The illiteracy rate is 37 percent, and the schools can accommodate only 41 percent of school age children. There is a need for 30,000 more teachers to provide the primary education of school-age children. Lack of sufficient funds for education led to a nationwide teach-

77. EBU Review, No. 72B, (Geneva), March, 1962, p. 26.

ers' strike in November 1964; more than 20,000 teachers had not been paid for over three months.[78]

Among Latin American countries, Colombia, a land mass the size of Texas and California combined, and with the largest television network in Latin America, had already made extensive use of educational television by 1965. Taking the Italian "Telescuola" project as an example, Colombia started, in 1961, experimenting with instructional television in the schools, a project which later expanded and included a full range of primary school subjects relayed, during the school hours, over the national network. The Colombian educational television is the greatest national commitment of resources in this field in Latin America. Five basic subjects are taught in primary schools and full attention is given to the adequate utilization of this broadcast in schools and to methodical evaluation and reporting of results. Schemes are also in hand for television coverage for adult literacy programs, for secondary education and for cultural extension services for adults.[79]

The United States contribution was not only $575,000 for supplying equipment for the project, including 1,500 receiving sets to be placed in rural schools, but also the assignment of ninety American Peace Corps volunteers who installed the sets and worked with Colombian technicians to write, direct and tape

---

78. New York Times, Educational Supplement, January 13, 1965.

79. B. P. Queenan, Senior Producer CETO, "Development in Educational Television in Latin America", CETO News, Center for Educational Television Overseas, No. 11, (London, June, 1966), p. 79.

the programs. By March 1964, about 470 sets had been installed and the taped courses reached approximately 100,000 children.[80]

Thus, the Peace Corps has been involved in one of the largest organized educational television projects in the world and its progress has "been steady and successful since its start in March, 1964.

> From a start of near zero, when school telecasts began in March 1964, Colombia Educational Television now reaches 400,000 primary school children and 6,500 teachers in eight departments of Colombia stretching from the center of the country at Bogota 700 miles north to the Caribbean. Included are the major population areas of the country. An additional 35,000 children and 700 teachers are viewing - "pirating" - the same telecasts in private schools which have purchased their own television sets. (According to the later report Colombian educational television in 1966 has reached almost one half million children and 1500 teachers in 9 departments of Colombia stretching from the center of the country in Bogota, 700 miles north to the Caribbean.)
>
> One thousand television receivers, out of a total purchase of 1,500 are presently installed in schools. The remainder are distributed in warehouses in key locations throughout the country for installation school-by-school, slowly, as each site is

---

80. New York Times, Educational Supplement, January 13, 1965.

carefully selected and satisfactory arrangements completed for the use and maintenance of each set.

All 1,500 television receivers will be in service by the completion of the present school year--that is, by the end of November 1966. By that time also over, well over, one-half million students will be receiving televised instruction.

Currently aired are: Mathematics - Grades 1 through 5 (the New Math); Language Arts - Grades 1 through 3 (Spanish, including grammar and written or oral expression); Natural Science - Grades 3 through 5; Social Sciences - Grades 4 and 5 (courses equivalent to U.S. courses in History and Geography; Music - Grades 1 and 2; and Teacher Training.

From a few limited hours of morning service at the beginning, educational television recently received responsibility for virtually all daytime television in Colombia, from 8 a.m. until 6 p.m., excepting only news broadcasts at mid-day. Thus, to the expanded school schedules will be added, next month, a pilot program in adult literacy. The full series for literacy has already been recorded on video tape. Also, program series in Public Health and Nutrition for adults, and Physical Education for adults and children are now in preparation and will be aired in the coming year.[81]

Peace Corps volunteers have been working on different areas of Colombia educational television particularly in production, programming, utilization and installation repair.

The use of television in education has linked the isolated schools that were spread throughout Colombia and thus has helped the meager educational resources of the country. Practical success of educational television in Colombia can be attributed to the fact that (l) the Colombian television network, as mentioned above, covers 85 percent of the country's area, and is the largest in Latin America; (2) the project is assisted considerably by foreign aid; (3) the native educators and broadcasters are realistic in their procedures and implementation of the plan.[82] Television, however, in spite of mitigating many problems, has a long way to go to solve the existing and emerging educational problems.

In Colombia there is great need for better educational facilities and excellent potential for providing these facilities through Colombia's Radiotelevisora Nacional. Almost half of Colombia's 14 million people are illiterate, the average elementary school teacher has little more than a fifth grade education and textbooks are scarcely seen outside urban areas. The Government of the Republic of Colombia owns and operates the largest television network in Latin America. 85 percent of the nation's population

---

81. Peace Corps, The Office of Radio and Television Programmes, Colombia ETV, a report, Washington, D. C., January, 1966, p. 1.

82. Dizard, p. 233.

and 94 percent of the public schools are within range of its transmitters. Before Peace Corps was invited to work on this educational television project, some attempts had been made to turn this medium into a valuable instrument of public education. But few schools had television reception and programming was limited primarily to schools in Bogota, the nation's capital. Now with Peace Corps assistance, full-fledged development of a nationwide educational television system is underway.[83]

The Colombian Government and business leaders agree that the Peace Corps project has been more than simply a demonstration. It serves as a pilot project which will be firmly incorporated into the national educational system. The Government is in accord with the broad purpose of the project, which is not escalation of the national educational level during the life of the pilot program, but rather the training of a cadre of Colombians professionally prepared to operate the facilities on their own when the Peace Corps project has come to an end. The original group of Peace Corps volunteers has fairly well resolved the initial details of program planning and production. Their first efforts have been concentrated on elementary school instruction. Colombian officials have estimated that these programs--including courses in 4th and 5th grade mathematics, natural science, geography and history--are reaching over 125,000 students. The future agenda includes not only expansion of programming at this level but also adult literacy programs, teacher training and courses in industrial skills.[84]

---

83. Peace Corps, Colombia Educational Television, Program Description, Peace Corps, Washington, D. C., 1965.

84. Ibid.

By the end of 1966, the Government of Colombia will have spent (pesos) $750,000 for 1964; $1,050,000 for 1965; and $1,000,000 for 1966 on its educational television service. The funds for educational television within the country are provided by three sources: (l) the Ministry of Education pays its 30 staff members who serve in the institute; (2) the Ministry of Communications, drawing on its own budget, on the sale of commercial radio and television time, and on the lease income derived from making new commercial television channels available to private investors; (3) the local departmental governments. The authorities of all income sources believe that educational television is off to a strong start and that the government intends to expand the programs, and to provide a prospective ground for Colombian leadership.

> Future Colombian leadership in the field is now being trained, on the scene. Recently an informal exchange agreement was concluded between Javeriana University in Bogota and the Radio and Television Institute, whereby the growing number of university students now being attracted to this new field in Colombia will be provided with a form of laboratory instruction in the studios and in the field as part of their college work, and at the same time Colombian technicians with courses of instruction in Communications Theory at the University. Such a step is of unusual significance, because university programs leading to careers in educational broadcasting, or any form of broadcasting for that matter, have been virtually unknown in Latin America

or anywhere else in the developing world until this time.[85]

Statements in such reports as have been cited above demonstrate that educational television in Colombia has been successful enough to persuade the educational authorities to employ the medium more extensively in Colombia's educational structure, although no specific research, to indicate the efficiency or deficiency of educational television, is yet available.

B. ARGENTINA

The Republic of Argentina, bounded by Bolivia on the north; Paraguay on the northeast; Brazil, Uruguay and the South Atlantic Ocean on the east; and Chile on the west, has a population of 21,762,000 people according to the United Nations estimate in 1963. Its capital, Buenos Aires, with a population of 3,799,000 people is the largest city in Latin America and 10th in the world. The language is Spanish and education is free, secular and compulsory.

In the Republic of Argentina, a considerable part of the national budget has been set aside to make education free from kindergarten to university. The 30,000 scholastic establishments are either state or privately controlled" in the proportion of 80 percent state establishments to 20 percent private. 35 percent of schol-

85. Ibid., p. 3.

ars are in Buenos Aires and its province "attended by 46 percent of the total scholastic population."[86]

Education in Argentina is provided on five levels:

(1) pre-primary education or kindergarten from 3 to 5 years, is under national, provincial or municipal jurisdiction. There are 2,500 kindergarten schools, 1600 state and 900 private schools attended by 120,000 pupils and taught by 7,200 teachers.

(2) primary education from 6 to 12 years, is divided into two groups: (a) normal school age, and (b) adults or children over school age. The 20,200 primary schools composed of 18,500 state and 1,700 private schools are attended by 3,200,000 pupils taught by 175,000 teachers.

(3) the Middle Education encompasses 6,500 institutions, 2,700 state and 3,800 private, attended by 1,000,000 pupils and requiring teaching personnel of 111,000.

(4) University Education which is rendered by 11 state (national or provincial) and 19 private universities. The 30 universities combined are made up of 190 faculties, and 200,000 students are taught by 10,500 lecturers.

---

86. Ing. Jose Finocchiaro, Director of Telescuela Education and Television in the Republic of Argentina, CETO News (London), No. 10, (March, 1966), p. 6.

(5) Higher Education is higher than the middle schools but not up to the level of University. The 260 higher educational institutions - 190 state and 70 private - are attended by 60,000 pupils and taught by 6,100 teaching staffs.

It is to these facts and circumstances that use of television in Argentina must in the first instance be related.

For years, from its inception in 1951, television in Argentina was limited to one state channel in Buenos Aires. Three of the existing six channels of Argentina were installed in Buenos Aires. In 1965 there were 21 open-circuit stations in operation, covering almost half of the Argentine territory. In addition to about 20 privately owned closed circuit channels, there are a number of educational closed circuit channels which, like those installed in the University of Buenos Aires and the Universdad del Salvador in Buenos Aires, are operating in different centers of higher education of the country.

The first live closed circuit transmission of the Faculty of Exact and Natural Sciences of the University of Buenos Aires that was viewed on receivers installed in a 70 meter clock tower of the Exhibition Hall of the City Council, took place in July 1963, and the recorded programs were transmitted from July to November

1964.[87] The Ford Foundation, from March 1961 to April 1965, has invested a total amount of $81,600 in the Faculty of Exact and Natural Sciences of the University of Buenos Aires, for experimental and demonstration projects to develop a group of educators and technicians expert in the methods and technology of Educational Television. The successful results of the project have attracted other Faculties which were distantly located, such as the Medical Faculty, to such an extent that they too are considering the use of educational television in their instructional programs.[88]

There are no open circuit channels exclusively for educational television. Educational programs, however, are carried on commercial stations. Broadcasting of an educational nature undertaken by well-meaning individuals, officials and private institutions, has been faced with problems and discouragements. Broadcasts for children and for schools have not always been supported by the directors of the channels whose main interest was to make money. "The four commercial channels broadcast up to 50 hours a day and these obligations exclude the possibility of broadcasts of an educative character at other times except in

---

87. Ricardo A. Romanelli, Closed Circuit Television in the University of Buenos Aires, E.T.V. Department, CETO News (London), No. 10, March, 1966, p. 13.

88. K. N. Rao and Robert S. Wickham, A Review of Educational Television in Latin America, report to the Ford Foundation, November 28, 1966, p. 2.

the morning for school children and in early afternoon for housewives."[89]

To remedy the above condition, the authorities of the Ministry of Education and the directors of the government broadcasting service made an agreement to bring into being a so-called telescuela program. Thus the first Telescuela Tecnica Argentina went on the air on the 15th of April 1963, broadcasting initially for three hours weekly. Telescuela, which is addressed to adult audiences, is devoted to the promotion and development of up-to-date technical education among men and fashion design among women in courses for which there is formal enrollment. There are almost 6,000 regular enrolled students in electronic mechanics for men and fashion design for women. But it also serves as a means of educational development by reaching sectors which are isolated and often deprived of other facilities, thus contributing to their cultural progress. Up to now results have been encouraging. The difficulties and pessimism of the beginning have been overcome and what to many people seemed impossible of achievement has made consistent progress month by month.

Argentina's educational television programs, like Colombia's, were inspired originally by the techniques of the Italian Telescuola. A series of high school courses on the state channel in Buenos Aires was inaugurated by the Ministry of Education in the spring of 1963. Telescuela Tecnica Argentina[90] was trying to

---

89. Carlos Alberto Duhoury, S.J., Technical School Television in Argentina, Television and Adult Education (Paris), No. 13, (May 1964), pp. 7-10.

90. Telescuela Tecnica Argentina is an educational establishment which comes under Consejo Nacional de Educacion (CONET). As regards its budget and its administration it is similar to any industrial school, but with reference to its educational work

help school drop-outs and those who live in areas where there were no high schools. These series attracted an enormous audience of different ages, half over thirty, some seventy years old or more.

The original three hours of weekly broadcasts offering a major in electrical mechanics for men and fashion design for women were later on supplemented by regular high school lessons in literature, English, mathematics and Spanish. Telescuela Tecnica Argentina had attracted an audience of 600,000 viewers by the end of 1963 and this encouraged the Ministry of Education to plan for making Telescuela a permanent feature of Argentine television.[91]

The agreement of 1965 between Consejo Nacional de Educacion Tecnica (CONET)[92] and Producione Argentinas de TV (PROARTEL), Channel 13, indicates the relationship of the educational system and the television companies. They agreed "... That, conscious of the responsibility which they have towards the community, the controllers of PROARTEL give up their transmis-

---

and the type of teaching provided it is completely different from other schools. This has resulted in a type of structure and organisation which are unique. As Telescuela was created to develop education using different visual aids, as well as television, begun in 1963, it has incorporated in 1965 teaching by Radiovision (Fonovision) and has also begun to make films for showing in schools. Telescuela, is assigned a share of the budget within CONET like any other school which comes under that Council. The monthly budget for paying salaries and working expenses is about one million pesos, of which about 97 percent goes on salaries for personnel and only three percent on working expenses.

91. Duhoura, loc. cit.

92. Technical education in Argentina comes under the National Council of Technical Education (Consejo Nacional de Educacion Tecnicas Conet), an autonomous body with facilities to nominate and replace personnel, to establish and close down schools, to arrange its own plans and programs, etc. It was created in 1958.

sion time and technical facilities for production, and the different private channels in the interior of the country give up their television air-time, for the broadcasting of Telescuela's programmes. For its part, CONET offers the service of Telescuela with a staff of specialists in teaching by television with three years experience in that field. This experience enables Telescuela to take to the television studio programmes which have been prepared and rehearsed ready to be recorded on video-tape. These programmes are prepared in accordance with a plan previously approved by CONET, following the educational and teaching patterns most appropriate to teaching by television and trying to utilise as much as possible all the resources which television provides."[93]

Educational programs constitute 10 percent of the total broadcasting hour and Telescuela lessons are recorded on video-tape, and thus are made available to other television channels throughout the country.

> For Telescuela's educational television operations, its staff of 70 teachers and technicians is divided into different sections according to the duties which they fulfill, teaching, technical or administrative.
>
> Each subject which is broadcast is prepared by a team made up of three teachers: a TV teacher who appears in front of the cameras and who is also a specialist in the subject; a TV teacher who can replace the regular teacher in front of the cameras and who collaborates with him in the prepa-

93. Duhoura, op. cit., p. 8.

ration of the programmes; a television assistant in charge of writing the script and checking it during rehearsals.

> The team has to prepare a plan of its lesson and a list of the necessary materials for its development, two weeks before the recording VTR or the live transmission, to give time to the different sections to prepare the illustrative material, captions, models, etc. Four days before the recording on video-tape a dry run is held and two days before recording a general rehearsal of the programme takes place in the replica of the studio which has been prepared for this purpose in the Telescuela building.
>
> So that full utility can be derived from the rehearsals, there is a team of teachers and actors in charge of the correction of errors in the presentation before the cameras, and of the re-design of the artistic side of the programme, if either of these steps is considered necessary.[94]

The lessons are summarized and published in the form of a booklet, or as separate lessons, and are given free to the registered pupils and available at cost price to those interested. To overcome the problem of one-way communication on television, reception centers have been organized.[95] Better use of television

---

94. Ibid, pp. 9-10.

95. Reception centers refer to the organized meeting places where the pupils can meet in front of the television set with a teacher or an instructor in charge of each center.

receiver, systematic completion of exercises and practice work, clarification of doubtful points, amplification of the important points, discussion and analysis of the subject matter, periodic examination for the pupils, and evaluation of the results, were among the various advantages that were derived from these reception centers.[96]

Among recent activities that have taken place in Argentina's educational television, one is the installation of a new channel for the audio visual department in La Plata and another in the Universidad de Tucuman. It is planning a major sociological study into the effects on the surrounding region of the introduction of television.[97]

C. BRAZIL

With a total population of 77,521,000 people, according to the United Nations estimate in 1963, and an area of 3,286,170 square miles, Brazil constitutes the largest nation in Latin America in population as well as in area. The country is bounded on the north by Venezuela and Dutch, British and French Guiana; on the east by the Atlantic Ocean; on the south by Uruguay, Argentina and Paraguay; on the west by Bolivia, Peru and Colombia. Primary education is free and compulsory. The language is Portuguese.[98]

Educational system in Brazil is a uniform and standardized system throughout the country, though it reaches a small percentage of the population. In spite of the rapid expansion of elementary education only about half the school age children

---

96. Finocchiaro, op. cit., p. 10.

97. Queenan, op. cit., p. 79.

98. World Almanac, 1965.

(between seven and fourteen) are attending schools. Elementary schools are maintained mostly by government. Only 10 percent of the schools are maintained by private organizations. Secondary education is almost impossible for the majority of Brazilians. It is mostly reserved for the elite. While the majority of elementary schools is supported either by state or municipal funds, the majority of secondary schools is supported privately by religious orders or private owners. At elementary level, 5,000 to 8,000 teachers are needed annually to maintain the status quo.[99]

Between 1950 and 1961, the population of Brazil increased by 40 percent and the number of pupils entering Brazilian schools almost doubled. The following figures indicate the growth in thousands:

	1950	**1961**	**Percentage of Increase**
Total Population	51,944	73,088	40
Primary Education	4,352	7,798	90
Secondary Education	540	1,302	160
Higher Education	49	102	108

([100*])

There are 32 official universities plus three Catholic universities and 486 others of high rank. Yet in 1965 48 percent of the adult population were illiterate. The utter inadequacy of its school system is attested to by the fact that in 1961 the federal law was

99. Charles Wagley, An Introduction to Brazil, (New York: Columbia University Press, 1963), pp. 204-231.

100*. The World Year Book of Education, 1965, (ed.) George Z. F. Bereday and Joseph A. Lauwerys, (England: Evans Brothers Limited, 1965), p. 358.

enacted by the terms of which all industrial, commercial and agricultural enterprises in Brazil employing more than 100 people were required to provide free elementary education to their employees and dependents. Provision for this requirement could be made by setting up their own schools or by contributing funds to public or private schools. This provided some additional funds and teaching services.

There had been some noticeable progress in the field of broadcasting as well. The number of broadcasting stations has increased from 367 to 803, between 1950 and 1961. The system is composed of government operation as well as of privately owned and operated enterprises.[101]

Developments in educational television date back to 1964 when Brazil (like Colombia, Argentina and over half a dozen other Latin American countries) instituted two courses to combat illiteracy, one elementary and the other more advanced. In this experimental course of TV escola 80 percent of pupils were successful compared to those in conventional courses where the ratio was 50 percent.[102]

Concern with television's educational potentialities is not, however, confined to government. Private individuals too have shown concern. Thus Joao Baptista do Amarol, chairman of seven Brazilian television stations, on his return from Italy, having been inspired by RAI, arranged for each of the seven stations he controlled to contribute one million cruzeiros (equivalent to

---

101. Ibid.

102. Alfredina de Paiva e Souza, professor at the Institute of Education, Rio de Janeiro, "Educational Television: A Brazilian Experiment", Television and Adult Education, (Paris, 1965).

$545) and 5 percent of its yearly income in addition to providing free time for educational programs to be broadcast between 6 a.m. and 6 p.m. daily.

Help from abroad has contributed to the belated but growing use of educational television. The Ford Foundation has invested $177,000 in Brazil, since 1964. The project for which the investment was made involved university level education of an experimental and demonstrational character encouraging experimentation with the new medium in teaching large university classes in teacher training. It was aimed at developing a group of educators and technicians expert in the methods and technology of educational television. The project was conceived of as a, model for the country and as a center for the development of videotapes and other materials for distribution nationally and for sharing with other countries in Latin America, as well as another method for accelerating curriculum reform and integration of university curricula.[103]

The problems faced in this, and in like projects elsewhere in Latin America, have been (1) lack of appreciation of the potentialities and the real role of the medium in education; (2) lack of technical competence in the university and the commercial networks resulting in delays in installation and efficient operation; (3) resistance by the educational establishment to the new medium; (4) lack of appreciation of the costs of such an operation and the consequent low budget provisions made for the programs;

---

103. S. Wickham and K. N. Rao, op. cit., p. 3.

(5) less than dynamic administration and promotion of the medium.[104]

The development of educational television in Brazil is both recent, piecemeal and rudimentary at the present time.

Some reasons have been cited for this tardy and snail's pace development. Nevertheless there seems reason to believe that a groundwork has been laid for constructive advances at all levels of service to education of which television is capable, as is evidenced by the following quotation:

> ... A steering committee under the supervision of the Conselno Nacional de Tele-Comraunicacoes in Rio de Janeiro has co-ordinated the activities of a wide number of groups interested in creating a national ETV system. It has been instrumental in securing the allocation of 103 frequencies for this purpose, and has assisted in the organization of several conferences and seminars. CONTEL has also assisted in the expansion of the pioneer ETV unit in San Paulo and the organization of a similar unit in Porto Alegre. The Committee has been encouraged by the success of these efforts to persevere in the aim of providing ETV facilities for the whole of Brazil.[105]

---

[104] Alfredina de Paiva e Souza, loc. cit.

[105] B. P. Queenan, op. cit., p. 80.

III. EDUCATIONAL TELEVISION IN AMERICAN SAMOA

American Samoa is a group of Polynesian islands (five of which are volcanic in origin and two are coral atolls) located approximately 2,300 miles south west of Hawaii. It is a territory administered since 1951 through the Department of the Interior. Prior to that time it served as a U.S. naval base and was governed by a naval officer. The main island, Tutuila, contains the port of Pago Pago and the seat of government. Approximately 18,000 of the total population of 22,000 live on its 43 square miles. Most of the remaining Samoans live on the three islands of the Manu's group, sixty miles to the east of Tutuila.

Before the use of television for schools on American Samoa, approximately 5,000 Samoan school age children received their elementary education in village schools of either open grass roofed buildings or somehow a better one - a wooden building with corrugated iron roofing. After finishing the sixth grade of elementary schools the pupil used to go on to one of several junior high schools which served the area. Previously admittance was on the basis of examination that deprived two-thirds of the children of attendance. Education was supposed to be in English but many teachers could not make themselves understood in the language. The instructional use of television in American Samoa took place on Monday, October 5th, 1964. This was the first introduction to television of any kind on that Island.[106]

---

106. Department of Education, Government of American Samoa, "Educational Television in American Samoa", a report by the Department of Education, CETO News (London), No. 8, (September, 1965), p. 12.

Shortly after his arrival as Governor of Samoa in May 1961, H. Rex Lee attempted to improve the educational system in that area. At his request, a new team was organized in the fall of 1961 by the National Association of Educational Broadcasters (NAEB) to study the entire educational system and suggest a feasible plan for improving the quantity and quality of education. The study group, after a thorough study, summarized the then existing problems as: (1) a lack of clearly-established goals for the schools; (2) a poorly-defined elementary curriculum; (3) a failure to successfully teach the fundamental skills; (4) the elementary school program provided no basis for an effective secondary education; (5) the method of teaching English was based upon class drill and rote memory and had no structural foundation; (6) methods of instruction emphasizing memorization with little attention given to the development of meaning and understanding; (7) a lack of appropriate instructional materials; (8) no audio-visual materials in the elementary or junior high schools; (9) overcrowding in the elementary and junior high school buildings; (1) a prevalence of poorly-educated and unprepared teachers; (11) inadequate supervision of instruction; and (12) ineffective teacher preparation and in-service training programs.[107]

A. ELEMENTARY EDUCATION

The television operation is designed in two phases: (1) over transmitters operating on VHF channels 2,4,5, at the elementary level of 1-6 and (2) over transmitters operating on VHF channels 8, 10, 12, augmenting instruction for the elementary grades and providing instruction for the secondary schools. The first phase

---

107. Peace Corps, "Cooperative Instruction by Television in American Samoa", an outline of the organization and procedures, (Washington, D.C., 1965), pp. 2-3.

started in October 1964 and preparatory plans for the second phase were envisaged for September 1965 using the same technique of direct instruction as was used in the first phase.

The first problem was to provide a multiple-channel television broadcasting system to reach all the projected schools with adequate signals. The system had to have sufficient height and power to radiate over about 70 miles of water to serve the villages and schools on the outlying islands and Western Samoa. The power of the transmitters and the angle of radiation of the antenna had to be low enough to be well received at different points on Tutuila and yet high enough to be picked up by translators on the Manna Islands.

The second stage in the development involved the problem of providing a television production center to feed the transmission system. The Television Center should contain the facilities for producing instructional materials, the work room, and research library, the master control room for the electronic transmission system, four television studios, supervisory offices, and technical shops.

> In the production center are four studios each equipped with two General Electric image orthicon cameras. A master control room houses all camera control units, switching equipment for the micro-wave, four film chains, and ten RCA television tape recorders. This permits us to achieve the ultimate of recording from all four studios while

simultaneously playing programs back on all six channels.[108]

The planning group recommended a complete and Immediate remodelling of the school system as well as the instructional program. The new system should provide educational opportunities for adults and children based on specific goals with new standards of achievement without sacrificing the desirable values of their Samoan culture.

The alternatives to the use of television were the following possibilities:

(1) Could an adequate number of new schools be constructed at a feasible cost and within a reasonable period of time?

(2) Would it be possible to employ several hundred qualified and dedicated stateside teachers?

(3) If such teachers could be employed in such numbers, would it be possible to provide adequate housing for them?

(4) Would the cost of recruiting such large numbers of stateside personnel, transporting them and their families to the Islands, and providing housing and other services be economically feasible?

(5) Would it be desirable to replace all Samoan school personnel with stateside personnel?[109]

The application of these alternatives meant a fantastic building program, overwhelming logistical problems, and the so-

---

108. "Educational Television in American Samoa", op. cit., p. 13.

109. "Cooperative Instruction by Television in the Schools of American Samoa, op. cit., p. 4.

ciological problem of dismissing about 300 Samoan teachers many of them with careers of more than 20 years of teaching. Television on the other hand could be the means of keeping and rehabilitating the teachers, providing instruction of a consistent and high quality, as well as being less expensive as compared to the alternative methods. Geographically, a study of the problem of electronic radiation indicated that television could be used for all the schools and could even be made available to Western Samoa, if desired. Thus the project for utilization of television was approved and the fund for carrying it out was appropriated by Congress and the NAEB was authorized to contract for equipment and stateside personnel.

Instructional television went on the air, for the first time, on Monday, October 5, 1964. The telecasting starts at 8 in the morning and ends at 3.30 p.m. The subjects in level 1 encompass language (native and English), mathematics, social studies; on level 2, science or health hygiene; and on the 3rd and 4th levels subjects such as planning and study, physical education, art and music study are added.[110]

The general operational activities of the staff of the Directorate of Educational Television are categorized into (1) instructional activities rendered by television studio teachers, assistant studio teachers, research teachers, publications personnel and librarians for the library of research material and visualaids; (2) the productional activities accomplished by the studio crews, producer directors, graphic artists and photographers; and (3)

110. "Educational Television in American Samoa", op. cit., pp. 15-17.

the engineering activities divided between studio and transmitter operators.[111]

Instruction is generally in English unless it is supplemented in native language by native teachers in order to ensure that the instruction is understood. (Oral English as a second language is taught to all levels. Reading and writing is taught to level 1, to make the pupils literate in their own language before attempting another.) Lessons are designed primarily for levels 1, 2 and 3 but many of the level 3 lessons are found applicable to level 4.[112] In-service telecasts for strengthening the Samoan classroom teacher, informing him of the content of upcoming lessons, acquainting him with new techniques, have also been scheduled.

Each school has a stateside principal in residence at the school site provided with a Samoan assistant principal. The basic instruction is presented by the television teacher; the classroom teacher, however, has the follow-up responsibilities, with the activities of worksheets provided by his television counterpart.

The Samoan television system, in addition to its daytime instructional programs, has evening programs in various fields such as English, Samoan languages, agriculture, public health and medicine, banking, taxes and certain trades. Other programming includes news, music, travelogues, feature-length movies and some entertainment movies from the United States. These

---

111. Operational aspects of educational television in American Samoa are the responsibility of a division of the Department of Education–one of the major divisions within the government of American Samoa. The division is composed of a Director of Educational Television and his staff members.

112. Grades 1 through 8 of the elementary education are now combined into a 4 level instructional program. Grades 1 and 2 were considered as level 1; grades 3 and 4 as level 2; grades 5 and 6 as level 3; and grades 7 and 8 as level 4.

programs are received by the villagers gathered in the consolidated schools each evening. The purpose here is to broaden the general information of Samoan dwellers.

Since improvement of the quality of education and raising the achievement level of the children are the two main objectives of elementary instruction, the core of the teaching is done by television and is more concerned with ideas, understanding and insights than with the memorization of relatively unrelated facts. The planning, presentation and reinforcement procedures are developed cooperatively as in "team teaching".

During the school year of 1964 to 1965, five consolidated elementary schools, in addition to the seven existing schools, were built and put into operation and the rest of the 26 proposed consolidated elementary schools are expected to be in operation during the ensuing school year. By the end of the first school year inclusion of televised instruction throughout the entire curriculum in the elementary schools was firmly established.[113]

B. SECONDARY EDUCATION

On October 1, 1965 the number of students enrolled in the three high schools, for grade 9 through 12, was 1695. In order to provide equal learning opportunities in all secondary grades in all high schools of American Samoa, to reorganize the traditional small classes into large ones, and to make possible a gradual build-up of Samoan instructional personnel and administrative staff, certain confusion and disadvantages, particularly in the period of transformation were inevitable.

---

113. "Cooperative Instruction by Television in the Schools of American Samoa, op. cit., p. 11.

In introducing the new system of instruction, the basic changes that affected the secondary school system were: (1) accommodation of large class groups that can be supervised, directed and motivated by a single teacher, with maximum flexibility, to accommodate areas for individual study and small special and remedial classes without impairing the efficiency of this instruction; (2) organization of instrumental teams designed to reorganize the staff by providing each high school with an American principal--who understands secondary curriculum and is experienced in cooperative television techniques as well--and qualified American supervisors of classroom instruction; (3) methods of instruction in which cooperative instruction by television--instead of single subject--matter teaching and comparatively self-contained classroom--is the core of instruction.

The main purpose of the television system of instruction, both in elementary as well as secondary level, has been summarized thus:

> The use of television in the educational system of American Samoa is designed to make the maximum use of a minimum number of American teaching and supervisory personnel during a developmental period in which as many Samoan teachers, supervisors, and technicians as possible will be used and upgraded to the extent of their abilities. It is also designed to provide maximum and equal learning opportunities to all Samoan children of American Samoa at a more rapid pace than has heretofore been accomplished.

> The end objective of the total effort is to provide the Samoan people with an adequate, modern educational system that will be within their social and economic resources.[114]

No formal research studies testing the efficacy of television's use are as yet forthcoming.

IV. EDUCATIONAL TELEVISION IN AFRICA

The I960 survey launched by UNESCO indicated that almost 85 percent of Africa's population was illiterate, compared to the world-wide average of 50 percent.

Seventeen million African children were receiving no schooling of any kind. Despite the need for vocational and technical training, there was not a single institute for training teachers in those subjects throughout tropical Africa.[115] The little improvement in those conditions that has been achieved since 1960, has brought no significant changes in the situation and Africa is still suffering seriously, from educational deficiencies.

A. NIGERIA

Nigeria, a country of 55,653,821 population (according to the United Nations estimate of 1963) is situated in Western Africa on the Gulf of Guinea between the Dahomey and Cameroon with the Niger Republic to the north and Chad to the north-east. It is composed of four distinct autonomous regions each with its own administration: northern, western, mid-western and eastern Nigeria in addition to the federal territory of Lagos, the capital

---

114. Ibid., pp. 21-22.

115. Dizard, op. cit., p. 235

of the country. This division has been applied to the educational pattern of the country as well.

Education in Nigeria is the responsibility of the four regional governments as well as the Federal Government, carried out through their respective Ministries of Education. The demand for education in Nigeria has been great, especially since World War II, and enrollment at primary schools has increased rapidly. Schools of higher education include University College at Ibadan, and the Nigerian College of Arts, Science, and Technology which has a division in each of the regions.

There are four independent radio and television organizations (1) the Nigerian Broadcasting Corporation; (2) the Broadcasting Company of Northern Nigeria; (3) the Western Nigeria Broadcasting Corporation; and (4) the Eastern Nigeria Broadcasting Corporation. The Nigerian television service which now operates under the Nigerian Broadcasting Corporation (NBC), is located in Lagos and its broadcasting can be received only in Lagos and its environments. Television programs for schools are broadcast only in the Northern and Western Nigeria through their Ministries of Education.[116]

The Western Nigerian television, the country's first television station, receives grants from the regional Ministry of Education. WN-TV started its first school broadcasting by offering two and one-half hours studio time to the Ministry of Education. The programs, encompassing education features, news and in-

116. Ernest Olaleye Leyimu, Head of School Broadcasting, Nigeria Broadcasting Corporation, "Proceeding of the Second International Conference of Broadcasting Organizations on Sound and Television School Broadcasting in Tokyo April 9-15, 1964, (Nippon Hoso Kyokai, 1965) p. 76.

terviews, English, geography and literature, were beamed for the secondary schools and teacher training colleges.

After independence in 1960, the Federal Government organized a National School Broadcasting Service for radio and television with the help of a Ford Foundation grant. This grant enabled Nigerian stations to make a modest start telecasting to classrooms programs of enrichment nature rather than direct teaching. Since 1963 the United States government is helping the Nigerian government by underwriting a considerable portion of the costs of a five-year program significantly expanding Nigerian educational television efforts. A curriculum of educational television lessons, developed by the United States educators and their Nigerian counterparts, was not carried out before 1965 because of financial problems and inability to form a broadcasting unit. The first broadcast for schools, after the establishment of the broadcasting unit, was a weekly lesson in science and English for elementary schools in the Lagos area. In the second term of 1965 the above subjects were broadcast twice in the morning and repeated by video tape for the afternoon session.[117]

By 1965 the educational programming of Nigeria Broadcasting Corporation (NBC) had been improved and following the advice of NBC school units, poor programs have been discarded and reliance on the stockpile has been discontinued, general improvement in educational content has been achieved, tape production services have been started, a magazine "Notes for

---

117. Marvin Bowman, "The First Year of Educational Television in Lagos, Nigeria," CETO News, (London), No. 13, December, 1966, p. 22.

Teachers" is published for each term and by 1966 a French handbook will accompany the radio course.[118]

The introduction of science into the primary school program of Northern Nigeria that combines theory and practice in its method of teaching, is one of the well grounded activities in utilizing television in education there.

> Writing, producing and presenting science programmes for primary school children in Northern Nigeria at this comparatively early stage of educational television development presents an extremely stimulating challenge. Flexibility of curricula and willingness on the part of the Ministry of Education to allow experiments in technique and approach, combined with the by now well-known advantages of television as a teaching medium, allow the television science teacher to exercise his imagination and ability to a degree unknown in almost all classrooms. However, a warning should be made that it is just this freedom of scope which can aggravate one of the major difficulties of educational television: the smooth welding of a television series into the fabric of the pupils' classroom course. A television lesson, no matter how interesting or well prepared, should not just form an isolated island of

118. Alan R. Beesley, NBC advisor, ETV in Nigeria, a report of the advisor of NBC School Unit to the Ford Foundation, (December, 1965).

completely new information and experience in the midst of normal classroom work.[119]

Different subject materials in the "junior sciences" in Northern Nigeria are treated accordingly with different methods and techniques. Thus, elementary general science is appropriate to be taught by fairly straight-forward teaching.

Nature study on the other hand requires a great deal of active participation by the pupils, so the object of this series is to act as a guide and stimulus to provide the student with just enough information to enable him to complete the necessary observation by himself. Here, Northern Nigeria, with its abundance of various plants and animal life, provides an excellent field of study.[120]

> Involvement of the pupil in a programme presents no difficulties whatsoever in Northern Nigeria. So complete is this involvement that the presenter of the programme must pause after his introductory "Good Morning" to allow the children to respond in their classrooms. Rhetorical questions have to be avoided because of the eagerness of the pupils to answer. In order to get the pupils involved in the subject rather than just the programme itself, each broadcast is "built-in" as just one part of the whole teaching approach to the subject. Classes were not expected to receive television lessons "cold".

---

119. A. B. Edington, Junior Science by Television in Northern Nigeria," CETO News, (London) No. 10, (March 1966), p.21.

120. Ibid., pp. 18-21.

An introductory session of roughly delineated topics--largely in the form of question and answer--was designed to stimulate students' awareness and make them receptive to the content material of the programme. Afterwards the programme should be supplemented by repeat demonstrations of the experiments by the teacher or the class, by other suggested experiments or investigations in the case of general science, by group project work in the case of health and hygiene or by each pupil completing a work-book in the case of nature study. Instructions regarding the supplementary work are issued to class teachers by the Ministry of Education together with relevant teachers' notes.[121]

There are, however, some serious problems, as attested to by a witness on the spot.

> ...Television is not a mass medium in Nigeria as it is in Europe. It is a luxury medium where sets are owned by Europeans working in the country, by the senior Nigerian civil servants and by Nigerian businessmen. Sets are expensive in Nigeria because of the high freight costs incurred in importing them and the very high customs duty.
>
> Television at this stage cannot help the problems of illiteracy or be much use in public enlightenment. Local authorities have set up communal viewing centres in Kano, Zaria and Kaduna, but

---

[121] Ibid., p. 19.

this can only scratch the surface of the problem. There are several ways of getting television sets to the masses, but each has its difficulties. Building or assembling sets in Nigeria would be a method of cheapening them, but with the high customs duty on importing components it is doubtful whether this could have any marked effect on the price. Assembly methods cannot be as efficient as they are in Europe and Japan, because sets would not be manufactured in such large quantities. It is also doubtful whether CIF prices (from European factories) can be much reduced.

The only sound method would seem to be the abolition of customs duty on television sets. Admittedly this would entail a loss of revenue for the Government, but with the limited number of sets being imported this loss would be small. On the other hand, if sets could be so cheapened as to sell in large numbers throughout Nigeria, then the commercial television stations could claim more advertising revenue, with a consequent reduction of government subsidies. In this way not only would the Government gain financially but, more important, television could become a mass medium. Great efforts could then be made to educate the population through television. This might further lessen the financial burden on the Government, as it would become unnecessary to keep the very expensive mobile film units on the roads. Also it would allow the Public Enlightenment departments to concentrate all their efforts on providing television programmes for the masses. What one mobile film unit can accomplish in two years, a television service can do in one evening, provided that sets are cheap enough to be as common as they are in Europe.

The abolition of customs duty in receivers, so enabling certain of them to sell at a little over &30, would probably increase the viewing audience by 500 percent in one to two years.

There is the added problem of lack of electric power in large areas of Northern Nigeria. This too will hold back the expansion of viewing, although with the Niger Dam Project this problem should ease as the years go by.[122]

B. UGANDA

Uganda is a country of 7,270,000 population (according to the United Nations estimate of 1964) which lies to the west of Kenya with Sudan on the north, the Republic of the Congo (Leopoldville) on the west, and Tanganyika, Victoria Nyanza, Lake Kioga and Lake George I on the south. Its territory includes part of Lake Albert and the Nile from Victoria Nyanza to Sudan. Its capital is Kampala. Uganda became independent within the Commonwealth October 9, 1962, and a republic October 9, 1963.

Television in Uganda was introduced in October 1963 as a division under the Ministry of Information, Broadcasting and

---

122. Leslie A. W. Diamond, Managing Director, Broadcasting Company of Northern Nigeria Ltd., "Bringing Radio and Television to Northern Nigeria", EBU Review (Geneva), 93 B., (September 1965), p. 29.

Tourism. In addition to its main transmitting station in Kampala, regional stations had begun to operate in Masaka, Mbale, Sorati and Lira. Another station in Mbarara is close to completion and those under planning are for Gulu, Masindi and Fort Portal. The 70 percent of the population presently within the coverage area of television stations will increase to 95 percent when the network is completed.

In July 1965, an educational school television service transmitting four times a week in the afternoons was set up.

> In the first phase of development, the ETV service has proved its worth to the schools system. Support material has been offered to teachers in Science, Music, Geography, English Language and Literature, Current Affairs, Arts and Handicrafts, Physical Education and a number of minor subjects. Various experiments and variations of technique and presentation were tested, and their results examined.[123]

Although radio at present, due to its being less expensive and more available, is more useful, television has a very promising prospect in the educational activities of Uganda. Given as a public service, educational programs form a significant portion of the weekly output of Uganda television.[124]

Presently some 50 schools are equipped to receive broadcasting and the number will perhaps increase when a plan for the

---

123. "Report on Uganda ETV Seminar", by Senior Producer, CETO, CETO News (London), No. 8, (September, 1965), p. 54.

124. Ibid.

extension of the network is completed. Teachers and specialists are in charge of the production of programs. They are invited on a part-time basis to write the script, prepare and present series in their own field of specialization.

An interesting innovation conceived of and executed in Uganda was the Kampala Seminar which was organized in Kampala by the Minister of Education of the Government of Uganda on techniques and principles of production of educational television programs in 1965. The Seminar consisted of two parts (1) the production techniques of the educational television staff and (2) utilization of broadcasts for schools and college principals. The course provided some useful data for similar courses in the future.

> In the first place, it is felt that short training courses of this kind have a definite value. They can be fitted into the normal breaks in the scholastic year and so can be attended by larger numbers of teachers and students working to rigid schedules. They can cover a wide range of topics, limited only by the level of preparation of the participants and the facilities available for practical work. In bringing together members of a unit who may have few other opportunities for discussion and interchange of ideas, they are invaluable for fostering teamwork and a common outlook. And, of course, they provide the basis for the standardisation of procedures and

techniques without which progress in ETV undertakings must remain erratic and haphazard.[125]

In addition to school broadcasts, there have also been programs of an informative and educative nature in the evening transmissions including two series of the "Arts in Africa", and "Science in Africa" presented by the Director of Extra-Mural Studies. Very little experimentation however has been done so far in directly educational programs for adults.

In the program of educational television development, in addition to educational broadcasting through open circuits, two closed circuits are being installed in the physiology department of Makerere University and in the Theological College of the Church of Uganda, at Mukono. "In both systems the intention is to integrate televised lessons into the normal syllabus of studies and to offer some training in television presentation in an academic context."[126]

C. ALGERIA

Algeria, with a population of 10,784,000 (according to the 1960 census) is a republic with socialist policies which became independent on July 3, 1962. It is situated in the North of Africa between Tunisia on the east and Morocco on the west. Its northern region extends inland about 350 miles and southward it extends into the Sahara Desert, merging into former French West Africa. Its capital is Algiers.

---

125. Ibid., pp. 54-55.

126. Ibid., p. 55

Education in Algeria follows the patterns laid down during the French administration. It is estimated that by early 1964 over one million children were receiving some education, but this represents less than half the school age sector of the population. In 1964 there were less than 500 Algerian graduate teachers. A large-scale campaign to combat adult illiteracy is being pursued with enthusiasm. Instruction is being given in some cases by students who have only recently left schools. There are two universities and a number of technical colleges. Many Algerians study in France. The numbers of students in the universities and colleges including the Faculty of Medicine were 6,888 in 1965.

Education (1963 Census)*		
Schools	**Establishments**	**Pupils**
State		
Public primary	2,195	750,564
Colleges d'Enseignement	150	14,623
Arab Religious schools	68	27,072
Technical	185	16,751
Agricultural	80	19,108
French Cultural Office		
Primary	300	(
Lycees	18	(91,512
Technical colleges	5	(
Private		
Primary	n.a.	11,970
Arab Religious schools	3	3,922
Lycees	n.a.	1,780
* incomplete		([127**])

[127] ** Source: The Middle East and North Africa 1965-66, 12th edition (London: Europe Publications Limited 1965), p. 106.

One of the serious problems in Algeria's educational structure has been the lack of qualified teachers. This problem obliged educators to recruit great numbers of supervisors in French and Arabic whose general educational level is hardly above the primary school certificate. To educate these young instructors, correspondence courses were organized through the national press. This venture was helped by radio on a small but growing scale. Radio had not been utilized specifically for schools in 1964.

> ... At present indeed, only courses in the French and Arabic languages are broadcast, intended respectively for instructors of levels 1, 2 and 3 which correspond to the first three years of secondary school. These radio courses are the living extension of those given in the press; they complete them without repetition and permit among other things the preparation of the students for follow-up studies of texts, the giving of dictation, and in a general way the guidance of their work by intelligent advice given by qualified teachers who thus aid them in making profitable use of the material provided in their textbooks and in their special weekly magazine.[128]

As far as television is concerned, though it covers a large part of the populated zone of the country, due to the high costs of receivers, so far, the Algerians have not been able to equip their

[128] Cherif Arbouz, Directeur de Centre d'Entrainement aux Methodes Audiovisuelles d'Alger Radiodiffusion-Television Algerienne, "School Radio and Television in Algeria", Proceedings of the Second International Conference of Broadcasting Organisations on Sound and Television School Broadcasting, op. cit., pp. 189-190.

school establishments with television sets. However, it has been used for literacy campaigns. 80 percent of the population is illiterate, and therefore Algerian television was mobilized for a literacy campaign and transmitted thirty minute courses in Arabic and French every evening.

A typical broadcast is given in the form of a reading lesson related to a language lesson and conducted in an adult class. The language lesson attempts to inculcate basic vocabulary elements; the reading lesson utilises the language elements acquired and exploits them, using a mixed method (half analytic, half synthetic)...But this experiment had great value as a test; first of all it showed the interest of the viewers in this type of programme; and it also gave a glimpse of a wider field and a more effective method of application."[129] The campaign, using psychologically inspiring slogans, was successful in awakening the people's consciousness of the "obscurity in which they had lived." In March, 1964, the number of adults passing with success the Primary Study-Certificate, ranged from 45 to 70 percent, while in the past the average was only 25 percent. In addition to this, programs on hygiene, child care and domestic science are offered in the form of demonstrations which introduce in the homes persuasive examples that only need following. These programs will be more extensive and refined in the near future. The number of programs pertaining to civic education, development of social consciousness, acceleration of the advancement of women, popularization of agricultural techniques will be increased.

---

[129] Ibid., pp. 191-192.

D. ETHIOPIA

With a population of 22,000,000 according to the Government's estimate in 1963, Ethiopia is a mountainous country in northeast Africa. It is bounded on the northeast by its state Eritrea which borders the Red Sea; on the east by French Somaliland, on the east, southeast and south by Somalia; on the south by Kenya; and on the west by the Sudan. There are a number of teacher training colleges including a university college in Addis Ababa, the capital of the country. The official language is Amharic though English is widely taught.

Education is expanding in Ethiopia. In the academic year of 1962-1963, there were 1,362 schools with 9,159 teachers and 304,138 students. A National Literacy Campaign, launched by the Government in 1963, involved 60,000 persons of all ages. There were 22 vocational and technical schools with 371 teachers and 4,800 students. Higher education is divided between the national university, Haile Selassie I University, in Addis Ababa with 1,042 students in 1962-1963 and study abroad comprising about 1,000 students.

Education (1962 - 1963)			
	Schools	**Teachers**	**Students**
Primary	1,278	7,664	288,496
Secondary	56	938	9,440
Special	22	371	5,162
University	1	186	1,041
			(130*)

Television in Ethiopia was inaugurated in 1964 under the management of Thomson Television International. From the time

130. Source: The Middle East and North Africa 1965-66, op. cit., p.122.

of its introduction in November, 1964, in Addis Ababa, television's educational value has been known and it was agreed upon that one of its main purposes should be education. Among the main problems in the way of the accomplishment of this goal were: (1) Can the Ministry of Education provide funds for a new and apparently expensive channel while schools lack more orthodox aids to education? (2) Where are the programs to come from? and (3) Who is going to provide the skilled staff to produce programs? In 1965, a scheme for educational television was rated and consequently two students were sent to attend the 5th CETO Conference in London. On their return, after preparing a number of reports on various aspects of educational television, the Ministry of Education asked them to inquire into the feasibility of extending educational service primarily to elementary schools throughout the empire of Ethiopia. Television was expected to be used as a means to assist rapid educational expansion and to provide "greater quality of educational opportunity in Ethiopia". The main problem was how television with its initially restricted audience might be helpful in providing the ground for implementation of these aims.

The initial program, however, envisaged launching programs for the secondary schools of Addis Ababa. The first transmission that was helped by CETO and the British Council went on the air at 11 a.m. on October 25, 1965, teaching physics, biology, English and geography. "We all thought it was marvelous, of course, because it didn't collapse and went out smack on the time."[131] The first session of educational transmissions which

---

[131] Brian Kirby, Television Officer of British Council, "Ethiopia School Television", CETO News, (London), No. 11, 1966, pp. 28-31. The author was one of two pio-

was composed of eight broadcasts weekly, was completed on December 24, 1965 "with encouraging results that provided an optimistic ground for the next session", that had to begin in March 1966.[132]

> We can look back on a successful start, for the reports received from the schools have been very encouraging and studio operations have been of a reasonable standard. We have been very encouraged by the work of technicians in the studio who have remained enthusiastic about our work. We have more personnel now some of whom are attending the ETV seminar in Cairo which began on February 5, 1966. The next semester will begin in March and we are optimistic.[133]

Thus the initial transmission of broadcasting for the schools of Addis Ababa that started in late autumn 1965 and was transmitted from the small two-camera studio in the municipality building dominating Addis Ababa was quite successful and gratifying as well.[134]

E. KENYA

Kenya, whose capital is Nairobi, is located in the east of Africa. It is bounded on the north by Ethiopia, on the east by So-

---

neers in educational television who went to London to attend CETO courses. The second was Abdurahman Mozayen.

132. Ibid., p. 30.

133. Ibid.

134. "Role of Educational Television in Ethiopia", CETO News, (London), No. 7, June, 1965, p. 50.

malia and the Indian Ocean, on the south and southwest by Tanganyika, and on the west by Uganda. It has a population of almost 9,000,000.

Radio and television both have been utilized in Kenya, under the Kenya Broadcasting Corporation. Television started in October 1962.

Kenya's public and private schools have places for only 1,250,000 pupils and only 10-15 percent of the pupils can be accommodated in secondary schools. There is no free education at the elementary or secondary level.[135] There are different languages and different races in Kenya.

School television has not been started so far. Indeed, television as such is in its literal infancy, as indicated by the fact that at the end of 1963, there were only 300 television receivers in Kenya with an average audience of 120 a set. However, the medium has programs for adult education. School broadcasting on radio that began in the spring of 1963 provides five hours of fifteen periods for 150,000 pupils of the primary and secondary schools, teaching English, history, geography, civics, economics and nature study. In addition to this radio has been providing ever-increasing adult education.[136]

By 1963 television had carried three main series of educational programs. The first one was on general health and hygiene for everybody, called "The Journey of Mr. Uhuru". It means the

---

135. Arthur John Raymer, Controller, Schools Broadcasting, Kenya Broadcasting Corporation, "Education for Adults in Kenya", Proceedings of the Second International Conference of Broadcasting Organisations on Sound and Television School Broadcasting, Op. Cit., pp. 54-61.

136. Ibid.

journey of Mr. Everyone in search of independence by which was meant freedom from poverty, freedom from ignorance and superstititions, freedom from sickness. The programs dealt with how to keep healthy and happy. There was a studio audience for this program and an African television doctor who was doing simple teaching with diagrams, animated strips and demonstrations, as well as answering the surprising and unexpected questions about everybody, men and women, children and adults, raised by the audience. It was planned and rehearsed. The second series was called, "Woman Makes the Home", in which a woman teacher with a studio audience was discussing and showing things related to women's activities such as child care, first aid and so forth. The third series of these educational programs was concerned with adult literacy itself--how to read and write. These programs were filmed under the direction of the Adult Literacy Center of Kenya, and were shown twice a week throughout the year.[137]

Funds for nearly all these new projects have to be raised outside Kenya. There is a great need for television and transistor radio receivers, a great demand for teaching materials. The average income is &20 to 40 a year and there are few good book stores and normally only in the towns, and suitable books in any language are scarce and expensive by the standard of the people's income. There is need not only for capital and recurrent expenditures but for research and follow-up in all these projects.

This problem is common to most countries that are working towards independence or nationhood on tight budgets. In Kenya the general view of radio and television is that of instru-

---

137. Ibid.

ments available to help overcome and root out the problems of ignorance, illiteracy and sickness that are the real obstacles to development, and to help in building a nature and well-informed nation. The main problems that set back Kenya's educational television development are (1) limitation of funds; (2) confinement of television coverage to a limited radius; and (3) scarcity of electricity. Steps have been taken, however, to reduce these problems. According to the educational authorities of Kenya, mass education and literacy are significantly important in building their nation, and that is why KBC is pressing on with its venture into adult education by both sound and television.

The transmittal coverage of television in Kenya did not go further than from 50 to 100 miles from Nairobi in 1963; this problem however has been remedied later on and the medium's coverage area has been considerably expanded due to the installation of new relay stations. As far as the level of educational programs is concerned the educators in Kenya were confronted with a problem of setting an average standard. Since 80 percent of the population are illiterate, the most conceivably reasonable standard adopted was that of the fourth elementary grade.[138]

F. LIBERIA

Liberia, with a population of 2,500,000 persons, as estimated by the Government, lies southeast of Guinea on the West coast of Africa, and is bounded on the west by Sierra Leone and on the east by the Ivory Coast Republic. It extends inland 75 to 190 miles, with a coast line on the South Atlantic of about 350 miles. The capital of Liberia is Monrovia and its official language

---

138. Ibid.

is English. Public schools are entirely maintained by the government. Others receive subsidies. There are nearly 4,000 schools, one university and two colleges. Founded as a Commonwealth 117 years ago, Liberia's economy is among the fastest growing in the world.[139]

> Outside Japan, Liberia's economic growth in recent years has been the world's fastest. The vast natural wealth of iron ore, rubber and diamonds is still in the early stages of exploitation. Although on a small scale as yet, a vigorous programme of industrialization is well established. With a dollar economy and highly-developed tastes Liberia is in step with western ideas on everything from architecture and agriculture to mass communication and mass education. It is far ahead of most countries on the African continent in its receptivity to progress.[140]

Prior to television, radio has been a well established medium in Liberia. The national radio service of the Liberian Broadcasting Corporation (ELBC) started on January 1, 1960, and by the time television was introduced, radio had 16 1/2 hours of daily transmission carried on 10,000 watt transmitters, on two wavebands broadcasting to 125,000 receivers. Radio is still the principal means of communication in Liberia.

Television arrived in 1963. In less than a fortnight after the equipment was landed at the Liberian Airport on December 12,

---

139. Douglas Grant, Sales Manager, Liberian Broadcasting Corporation, "Television in Liberia- a Sturdy Newcomer", EBU Review (Geneva), No. 90 B, March, 1965, p. 21.

140. Ibid.

1963, the station was ready for operation, and the Liberians were able to see and hear their president inaugurating television " as a fact in yet another African country".[141]

In 1965, after the establishment of regular programs, the ELTV station of the Liberian Broadcasting Corporation had extended its services through 2,000 receivers to approximately 14,000 people. It transmits over 3 1/2 hours of programs every evening of which 2 percent are live-from a 500 watt nominal transmitter installed on the topnof the Ducor International Hotel, 400 feet above the city and covers a population of approximately 200,000 people in the heart of Monrovia.[142]

The contents of the Liberian television run from light programs such as "Teenage Dance Time" and entertainments, to educational programs in the women's world, and "Be My Guest" that makes the audience familiar with political and social leaders. Television has not been utilized for schools or formal education (except French lessons). However, the use of the medium for educational purposes is being considered and planned. Although receivers are still limited to a minority who can afford them, the availability of receivers on rental basis is apparently a beneficial way of making television accessible to those who cannot afford to buy receivers.

> The Liberian Television Service already carries French lessons by television and plans an elaborate educational service. Not only will these be of value to schools, but also to many adult set owners

---

[141] Ibid

[142] Ibid., p. 22.

and those who rent their receivers from Rediffusion's subsidiary company in Liberia. At $2.00 per week, a rented television set provides the cheapest and best entertainment available. With sets accessible to nearly all but those in the lowest income bracket, it may well be that the Liberian Government will lead the way with education by television- without fanfares, without enormous expenditure and without high-powered experts- simply by doing it, much as they have done everything else.[143]

G. UNITED ARAB REPUBLIC (U.A.R.)

The United Arab Republic occupies the northeast corner of Africa on the Mediterranean sea. On the east lies Israel and 1,200 miles of Red Sea separating Egypt from Saudi Arabia. Libya is on the west and Sudan on the south. Its population according to the United Nations estimate of 1962 is 27,303,000 and its capital city, Cairo.

Education is compulsory for all children beginning at age 7 and free through high school The four main universities are Cairo University, Alexandria University, Heliopolis University, and Assiut University. In addition to this, there is a seat of Moslem learning in the University of Al-Azhar in Cairo founded in 968 A.D. Arabic is the official language. There are several American centers including American University, Cairo; American College for Girls for all grades through junior college; Cairo American College, private grade and high school, and others.

Education in U.A.R. (1962 - 1963)

143. Ibid.

	Schools	Teachers	Pupils
Primary	7,374	66,896	2,909,996
Intermediate			
General	955	9,591	349,661
Technical	142	1,742	49,667
Secondary			
General	239	4,236	134,047
Technical	104	2,655	79,639
Vocational	59	883	23,141
Teacher Training	41	n.a.	23,872
University	4	n.a.	98,537
			(144*)

In addition to the aforementioned local students, there are a considerable number of foreign students studying in the United Arab Republic's educational institutions.

Foreign Students in U.A.R. Universities (1962-1963)	
	Students
Arab World	14,477
Africa	798
Asia	480
Europe	6,159
America	418
Others	187
Total	22,514 (145*)

Prior to the revolution of July 23, 1952, in which the monarchy of King Farouk was overthrown, there was no television in

144. Source: The Middle East and North Africa, op. cit., p. 225

145. Ibid.

Egypt. There were only two radio transmitters, broadcasting 22 hours a day. Under the dynamic leadership of Nasser and his regime, in less than 6 years the United Arab Republic radio capabilities had been boosted to 11 transmission centers, broadcasting 198 hours daily and a new shortwave radio service broadcasting 130 hours a day in 19 languages.[146] And had it not been disrupted by the Suez crisis of 1956, television would have been introduced in the country long before 1960. These early plans, however, were revived by 1958 and the first television transmissions were made from Cairo in 1960, right on the eighth anniversary of the revolution.

> The U.A.R.'s television system is undoubtedly the most extensive and probably the most effective of its kind in any of the newly developing countries of Asia and Africa. Almost from the beginning, Cairo television and its coaxial links reached most of the heavily-populated areas of the Nile River Basin. In 1966, the completion of a microwave link with Aswan, three hundred miles south of the capital, will bring U.A.R. television within range of most of the nation's populace. The number of sets in 1965 was an estimated half million, serving an audience of over three million persons. While this is largely an urban audience, the U.A.R. government has made special efforts to make television sets available in rural areas for mass viewing. In 1965, the

---

146. The first president of Egypt after the success of the revolution in 1952, was General Najeb. Gamal Abdel Nasser became directly in charge of the new regime in April, 1954.

government announced that its television receiver factory had begun production of a battery-operated transistor set primarily for use in villages without electricity. At the same time, it declared that it would distribute two million of these sets to such villages during the first year of production.[147]

Incidentally, every Arab State except Yemen has television facilities serving rapidly expanding audiences. Television is already second to radio in influence and popularity. The United Arab Republic's programs include entertainment, education, Arabic songs and dramas, as well as American cowboys, detective stories and comedies. The United Arab Republic's television systems are under the guidance of the Ministry of Culture and National Guidance. Television signals can be received virtually throughout the United Arab Republic offering the audience a choice of three programs. Channel 9, the principal Cultural Channel, is, however, available in Cairo only. The existing six studios transmit a daily schedule of 25 to 30 hours to about 500,000 receivers.

> Unlike television in most developing countries, the U.A.R. program pattern is heavily weighted in favour of local shows. A 1963 survey showed that 78 percent of its programs were produced within the country. (In many other developing countries, the ratio would be almost reversed, with the bulk of programming consting of American and other Western filmed serials). Prom the salons of Cairo's exclusive Garden City district to waterfront

147. Dizard, op. cit., pp. 147-148.

teahouses at Port Said, television is rapidly taking its place as Egypt's most important entertainment medium.[148]

The Ministry of Culture and National Guidance in its outlining of broadcasting policy of 1960, in addition to the medium's responsibility for social, political and spiritual programs, has stressed the medium's role in education and enlightenment. The policy's highlights emphasizing educational activities are as follows:

(1) Raise the standard of the arts.

(2) Inform the people of the great achievements of human civilization.

(3) Give due consideration to vocational and craft programs.

(4) Combat harmful habits and traditions handed down by past generations, deal with social problems and call for closer adherence to spiritual and moral values.

(5) Give prominence to the role played by women in the progress of the family and society.

(6) Disseminate sports consciousness and the athletic spirit among the rising generation.

(7) Encourage new talents in the field of thought and creation.

---

148. Ibid., p. 148, as quoted by the author from "Television in Egypt" by Bahie Nassr, Broadcasting, June 10, 1963.

(8) Create the group spirit between individuals and groups.[149]

The enlightenment programs of the United Arab Republic's television system, aside from their entertainment, have two main purposes, (1) to function as an instrument for social education on subjects such as health, sanitation, home economics and school television. The third channel, a channel reserved for educational purposes, includes a plan for broadcasting instructional programs for the schools, and has been assisted by the U.S.A. through two audio visual specialists in 1963, and the specialists' draft for a five-year program for school television is slowly being implemented. (2) to perform its responsibilities as a mass medium of communication (as the ruling regime sees it) supporting the political principles of the country and disseminating the news for new national and international achievements and progress.

Ever since the 1952 revolution, one of the main objectives of the United Arab Republic has been the extension and improvement of education and the general enlightenment of the people. To achieve this goal television has been extensively used for mass education, and literacy campaigns. Thus in August 1963 special experiments were undertaken in utilizing television for literacy teaching.

> At this stage educational researchers, as well as literacy and television specialists, were brought in to plan the experiment, which it was hoped would yield answers to many important questions to do with the organization of such a project, as well as

---

149. Ibid., pp. 148-149.

> the ultimate question of the effectiveness of television in literacy teaching. Matched control and experimental groups were established and the television teachers were carefully selected. The project was publicized and field work was planned to assist the motivation of the students and to help maintain interest throughout the course. Finally, the teachers using the programmes were given two weeks' training in handling them in the classroom. The course itself ran from October 1963 until June 1964 and consisted of ninety-five lessons. At the end of this time it must have been disappointing to be able to report only a partial success.[150]

In order to make plans for utilizing the existing educational facilities to meet the pressing needs for education, and to emphasize the idea of Arab unity, the United Arab Republic became the host country for two seminars about educational television that were held in Cairo. The first Seminar took place in November 1964. Ten Arab countries were represented. Held under the auspices of UNESCO, this Seminar was particularly focused on future planning rather than on past achievements. At that time some of the Arabic countries such as Lebanon, Iraq and, of course, the United Arab Republic had already utilized television and others were either in the process of planning or in the first month of transmission. All the countries that had television systems, included some educational programs, ranging from broadly "cultural" or "educational" to such highly specialized areas as literacy

150. Geoffrey Hall, "The Cairo Seminar on Educational Television for Arab States", CETO News (London), No. 7, June 1965, p. 29.

teaching and the training of television servicing engineers.[151] The conference participants agreed that first priority should be given to the eradication of illiteracy which in some cases rises to a level of at least 70 percent.

For its second Seminar, UAR-TV invited CETO to organize a course in Cairo's TV Institute to train course members in the practical application of television in the field of education, particularly the medium's use in schools. Eleven members from Ethiopia, Jordan, Iraq, Morocco, Palestine, Saudi Arabia and Sudan were invited and took full part in the course. The first week of the course was mostly devoted to lectures in which almost every aspect of operating an educational television system was discussed. In the second week, the theoretical background of audio-visual education was given in a series of six lectures. Following the first two weeks, preparation for the first short training in production started.

> Course members were divided into teams, each containing both educators and television personnel. A director was appointed from each team, and the others were given studio jobs. Subjects of an educational type were allocated, each to last five minutes...The five-minute exercises were an important prelude to the twenty-minute programmes produced later. They gave practical experience to the TV director and educators in working together, and gave the studio staff the opportunity to accustom themselves to the particular requirements of ETV.

---

151. Ibid., pp. 28-29.

> The fact that the programmes were not uniformly successful was no bad thing as many lessons could be drawn from them; it was in fact noticeable that owing to improved teamwork the later productions were generally better than the early ones.[152]

Simultaneously with the progress of the course preparation for experimental school transmissions had been proceeding. It was decided that the most suitable level for the programs was first-year Senior Secondary, and six boys' schools and four girls' schools were selected.

An engineer and a member of the audiovisual department had to visit the schools, before transmissions, to check viewing arrangements with the school authorities. In order to emphasize the importance of contact between the television producer, director and schools, the teachers were asked to give their assessments of the value of the broadcasts.

In addition, three workshops, one for principals and senior teachers, and two for classroom teachers in the schools that were going to receive the experimental lessons, were held.

The screened programs aroused great interest, the teachers adapting themselves readily to television and leaving no doubt that when the problem of production and organization was solved there would be a basis in the schools for a national service.[153]

---

152. "Cairo ETV Seminar and Schools", Broadcasting Experiment February-March, 1966, Information and Research Division, CETO News, (London), No. 11, June, 1966, p. 68.

153. Ibid., p. 70.

V. EDUCATIONAL TELEVISION IN ASIA

Although Asian countries have made considerable progress in the last two decades, it has not been sufficiently effective to eliminate the current and emerging problems of their low level of national income and consumption, high rate of illiteracy, and low standard of health. If the human factor is one of the most important elements in development of societies and if Asia is the most densely populated continent in the world, then one reason for all their backwardness is lack of the motivation and education that are necessary for sufficient productivity. Motivation and education are prerequisites for the improvement of that productivity that ultimately leads to a better standard of living, economic development and prosperity.

> Unless the quality of its productivity is increased all the material investment that may be made will prove ineffective. We have to transform the people of Asia if we are to transform Asia.[154]

It is the motivated people who are receptive to education and consequently contribute effectively to the development of their society. This process of motivating people has proved feasible through the introduction of new technology--particularly television. Through the persuasive capabilities of television we can, hopefully, motivate, and through its educational potentialities we can provide education to all, and subsequently through the full use of this device we can, who knows, transform a nation.

---

154. "Meeting on Broadcasting in the Service of Education and Development in Asia", Bangkok, 16-23 May, 1966, UNESCO/MC/53, Paris, August 1966, p. 5.

A. INDIA

According to the United Nations estimate of 1962, the population is 449,381,000. The capital of India is New Delhi. The country extends south into the Indian Ocean, with Cape Comorin at its extreme tip, and Ceylon southeast across the Gulf of Mannar; the Bay of Bengal is to the east and the Arabian Sea to the west. Pakistan borders India on the northwest.

The constitution provides for free, compulsory education through age 14. The educational system of India is composed of three main levels, elementary, secondary, and university education. Elementary education encompasses two levels of schooling: Junior Basic Schools in which a child of 6 starts from grade I and completes grade V at the age of 11; Senior Basic Schools, starts at grade VI and ends at grade VIII. Secondary level of education starts after the completion of elementary level to grade XI. 11 years of study are required before entering the university. Technical curricula are state supported. There are now 46 universities, 1,946 colleges and 27 research institutes. There are 14 main languages, 12 originating from Sanskrit, with several hundred different variations in dialect. The official language is Hindi.

Pressing demands for more and better education on one side, and shortage of teachers, textbooks, buildings and other educational facilities on the other are among important factors that lead India to seek possibilities for a practical solution of a seemingly insoluble problem. Radio had already been serving as an in and out of school educational medium, and the next step involved the use of television.

Broadcasting in India is a state controlled public service system and All India Radio (AIR) with its dozens of regional stations and auxiliary centers, is the only broadcasting authority in the Country.[155] The comparatively long hesitation of All India Radio in introducing television has been attributed in large measure to Premier Nehru's attitude towards the medium. He was of the opinion that "television is a luxury that a nation such as India, great parts of which are still threatened by the pangs of hunger and extreme misery, has no right to offer to more privileged classes."[156]

In spite of this the government authorities were well aware that television can be a tremendous force for dissemination of education among the masses, especially in rural areas, and a start was made, in 1959, with an experimental television service within a narrowly confined terrain comprising the urban and rural area of Delhi and the population within a radius of 12-15 miles. A weekly one hour program was broadcast. Its contents were of an over all educational character planned for community viewing by adults, and for classroom viewing by schools.

Two major television projects, one in collaboration with UNESCO and the other with the Ford Foundation, were developed. The UNESCO project of Social Education Through Television started in Delhi toward the end of 1960 and was completed in May, 1961. The project consisted of a series of broadcasts to 66 community centers, each with active members of 15 to 20 and casual viewers of 100 to 150 persons. Twenty of these centers were chosen for study of the effects of television on adults.

---

155. V. K. Naryana Menon, Director General of All India Radio, "India in the Approach of a Secular State, EBU Review, (Geneva), 97 B, May 1966, p. 68.

156. "In India", Television and Adult Education, (Paris), No. 13, May 1964, pp. 28-29.

The content of twenty weekly programs, to be offered over five months, consisted of topics related to specific problems in the community and aimed at inculcating good citizenship.

> The main purpose of the inquiry was to assess the usefulness of television for social education. This was to be gauged by measuring the shifts in information, attitudes, and behaviour brought about in the 418 members of the 20 tele-clubs in the sample, as a result of viewing 20 special telecasts on citizenship and participating in post-viewing discussion.[157]

The research findings disclosed that significant shifts in the three aforementioned areas occurred. The members of the experimental group increased their main scores on information, attitude and behaviour questions considerably. The increase in a control group was on a smaller scale as compared with that of the experimental group.[158]

Being greatly encouraged by the result of this experiment the Indian government not only continued the project but expanded it to such an extent that by 1964 there were 180 tele-clubs in Delhi, with expectation of 200 tele-clubs with 4,000 active members by the end of the same year.[159] Tele-clubs have now become

---

157. "Social Education Through Television", an All India Radio UNESCO pilot project, Reports and Papers on Mass Communication, No. 38, UNESCO, (Paris), 1963, p. 35.

158. Ibid.

159. Ramesh Chander, "Sound and Television in the Fundamental Instruction of Adults-Citizenship through Television, All India Radio", Proceeding of the Second International Conference of Broadcasting Organizations on Sound and Television, op. cit., p. 444.

an important factor in the life of their members and there is a desire to add more, if television receivers can be provided.[160]

The second project alluded to above was for school television and was launched on October 23, 1961 with the help of the Ford Foundation. Its primary concern was the teaching of science (physics and chemistry) to the higher secondary school classes. Eight lessons of twenty minutes each were broadcast weekly and were repeated in the afternoon for the second shift students. English and Hindi were also included as subjects in these lessons. These programs were organized on the basis of team teaching.

The number of lessons was increased to 13 in 1964 and to 16 in 1965 covering (in addition to science for higher secondary) English for 6th, 7th and 8th class and social studies for 8th class only. Besides this, one program for teacher training in English once a month and on science twice in three months has been embarked upon.

> The lessons are directly related to the curriculum and make full use of the visual-film strips, photo-still, maps, pictures and laboratory experiments. The project has been undertaken with the cooperation of the Education Department of the Delhi Administration. All higher secondary schools will be equipped with television sets. In October 1961 a beginning was made with 146 schools and the project has covered 252 schools by now and will ultimately cover nearly 60,000 students. Early in 1961, the Government of India entered into a four-

160. Ibid.

year agreement with the Ford Foundation for starting regular school television programmes for students of all higher secondary schools in Delhi. The agreement provided for the installation of about 600 television sets in the schools during the four year period, the supply of some necessary technical equipment to AIR, the training of AIR personnel and the deputation of educational TV consultants from U.S.A. to India.[161]

The more specific objectives of television's class-room use are suggested by the project's education officer as follows:

> Effort is made to make the TV presentation a model presentation, acting as a guide for the order of presentation of structure, for the method of presentation and for pronunciation. (In this sense the lesson incidentally serves as a means of in-service training for teachers.) The TV lesson, by going beyond the immediate range of classroom possibilities and illustrating a teaching point in a real life situation, can enrich and enlarge the child's experience by taking him out of the classroom. The proper exploitation of the TV medium by means of films, studio sets, animated captions, puppets, models, toys and illustrations, etc., can take the child into the country, into the city, to the seaside or into a world of fantasy and make-believe. It thus takes into account the psychological attitude of the

---

161. "Television in India", Indian and Foreign Review, Vol. 2, No. 18, July 1, 1965, p. 13.

viewers to TV in general and language teaching in particular.[162]

Television's school programs tend to create passivity unless caution is exercised in adopting the most suitable pattern of presentation to attract and maintain the interest and participation of the students throughout the program.

> The possibilities include: (1) televising an actual classroom; (2) televising a teacher teaching in vacuo; (3) using exclusively cartoons, films and playlets, and (4) televising a team presenting a simple dramatic situation which illustrates a teaching point with a teacher commenting on the underlying teaching point.
>
> The first possibility is uninteresting because it is dull to watch classroom drill where there is no opportunity for viewer participation; and the second possibility gives little opportunity for seeing language in action; as for the third, few suitable films and cartoons are available. The fourth possibility (i.e. the situational method) is the most suitable, as it demonstrates language in action and, for variety, can introduce elements of the other three methods.[163]

To help the classroom teacher a guide sheet was prepared for each TV lesson, defining the object and scope of the lesson,

---

[162] J. G. Mills, TV Education Officer, "Delhi ELT/TV Project", CETO News, (London), No. 7, June, 1965, p. 7.

[163] J. G. Mills, op. cit., p. 8.

indicating the vocabulary used, outlining the experiments made in the lesson, suggesting points for preparation and follow up in the classrooms and including a list of reference books. Out of 366 government schools in Delhi, 238 have television sets with programs viewed by approximately 229, 000 students. At the beginning of each TV lesson the classroom teacher prepares the students for five minutes, and the lessons generally take 15-20 minutes and are immediately followed up for a further 15-20 minutes.

The above experiences have been encouraging to the extent that during the Fourth Five Year Plan a proposal for the expansion of the number of TV centers to Bombay, Madras, Calcutta, and possibly to Kanpur, was made. This proposal was in addition to the one for expanding the centers at Delhi. The stations for these centers will have a range of 80-100 KM. like those developed in Delhi, covering the rural areas close to these cities by satellite stations, auxiliary stations or by co-axial cable. In addition to arrangements for the importation of over 9,000 receivers from various countries on rupee payment, the Central Electronics Engineering Institute at Pilani (India) started production in 1965, and is expected to produce almost 1,000 sets a year.

> The Delhi TV Pilot Project is still in an experimental stage and many of the problems have yet to be overcome...There will be better coordination (in future) between school syllabus, the school

textbooks, and the TV lessons, and the examination system...As far as the children are concerned, a visit to school with TV facilities to observe a TV viewing class evidences the children's active participation in, obvious enjoyment of and benefit from the English by television lessons. (As far as viewers research,) the effectiveness of the lessons can be only judged imperfectly by visits to schools, through discussions with teachers and children, at seminars and group meetings of teachers.[164]

B. PHILIPPINES

The Republic of Philippines, a country with a population of 30,331,000, as estimated by the United Nations in 1963, lies between 21° 20' and 4° 30' North Latitude and between 116° 55' and 126° 34' East Longitude. The country is composed of 7,100 islands extending 1,150 miles North to South and 682 East to West.

Education in the Philippines is free, secular and coeducational. In 1953, over 4.5 million pupils were enrolled in about 31,000 schools and institutions of higher learning including the University of Manila which had nearly 35,000 students.[165] Shortage of qualified teachers and other educational aids and facilities are among urgent educational needs, not only in formal schooling but also in informal teaching such as the campaign against illiteracy which encompasses about 35 percent of the Philippines'

---

164. Ibid., p. 12.

165. "The New International Year Book: A Compendium of the World's Affairs for the Year 1965" (New York: Funk and Wagnalls Company, Inc., 1966) p. 368.

total population. To overcome these problems, the educational authorities have utilized the broadcasting system, both radio and television, in the service of education.

Educational television, though started as a joint enterprise between the Bureau of Public Schools and the Philippines Broadcasting Service, is an example of a "soundly organized" closed circuit system extending beyond limitations of the closed circuit link because of the need in the broader community of schools. The closed circuit system was set up by the educational authorities and its facilities are used for recording educational programs to be transmitted over commercial or national network.

The way for the introduction of educational television in Philippines was paved by its predecessor, the medium of radio. Due to the encouraging results of radio's use, introduction of televised courses for the public secondary schools came under study. Thus, between 1960 and 1963 personnel from the "Radio Education Unit" and the Philippine Broadcasting Service were sent abroad to study the application and uses of sound broadcasting and television in education. When the government-owned television channel, channel 10, went into operation in January 1961, the Bureau of Public Schools started a first experimental television series which was a course on speech improvement, entitled "Improve Your English", for students of the secondary schools. This course was followed by students in the third and fourth years of the secondary schools. An additional course was offered in Physics for fourth year students. These were weekly half-hour live telecasts for enrichment. In addition to these enrichment programs, a half-hour live program for adult and community information was also broadcast. "Education on TV", a half-hour program series, was broadcast three times a week on channel 9, featuring outstanding representatives of Manila University. This program provided an opportunity for university professors to appear for the first time on the screen.

In the summer of 1961, television's educational use was extended to contribute to higher education. A course, "Physics for the Atomic Age" was started as a project of the National Science Development Board, offering televised courses in introductory college physics to ten participating colleges in Greater Manila and its immediate surroundings.[166] Though the course was offered for credit, there were many non-credit viewers as well, composed mainly of physics teachers in secondary public schools and their students. "This first venture in educational television in the Philippines was made possible through cooperation of a number of different groups and individuals all of whom were convinced of the need of such a project."[167] This need for cooperation that avoids wasteful duplication and secures quality production led several Philippines educators to pool their resources and efforts for the creation of an organization to take charge of instructional television. Thus the Metropolitan Television Association (META) came into being in early 1963 and was administered by the Center for Educational Television. This Center is one of the most extensive training centers for educational television in developing countries.

---

166. Vitaliano Bernardino, "Director of Public Schools Department of Education, "A Brief History of the Development of Instructional Television in the Philippines", Proceedings of the Second International Conference of Broadcasting Organizations on Sound and Television School Broadcasting", op.cit., p. 149.

167. Ibid.

> The META has initiated an instructional television programme for secondary schools in the Greater Manila area called "the National Programme of Instruction by Television in Secondary Schools". The project is scheduled to offer televised lessons in daytime via the broadcast facilities of commercial television stations to selected high schools in the Greater Manila area and is to start in August 1964.[168]

Two television courses were planned for the school year 1964-65--a course on physics for the fourth year and English for the first year. The physics course was produced on kinescope to enable reshowing of the film at different hours to meet the demands of varied school schedules and for eventual distribution to other parts of the country. A Seminar-Workshop in Instructional Television has been scheduled for the summer of the present school personnel who will be involved in the project.

The National Science Television Board, on the other side, had been conducting television courses on Modern Chemistry during the school year of 1962-63 in which 17 universities and colleges of the Greater Manila area are participating.

In 1964 the Ateno de Manila University was helped by the Ford Foundation and the local business firms to create a center to carry out two projects: (1) training educators in the use of the medium as well as other audio-visual devices, and (2) developing a "phased program" for the broadcasting of instructional material to the schools throughout the country. For the first project the

168. Ibid., p. 150.

university opened a new section in its Department of Education, at a graduate level, to train educators in this new field. "Classroom theories in this new field are tested by the students through practical experience at the Ateneo's Center for Educational Television." The Center is well equipped and can train not only the students of the university, but administrators, studio instructors, and classroom teachers of other institutions as well.

> The Ateneo Center has two other ETV projects currently under way, The first is its own closed circuit teaching system for the Ateneo and the adjoining Maryknoll College. Programs for this internal ETV system can be produced at the center and "piped" into the nearby classrooms. During its early experimental phase the center is evaluating the range and types of instruction best suited for university conditions.[169]

The Ateneo has also contributed to the development of instructional television at the elementary and secondary level. Significant has been the extension of in-school instruction outside the Ateneo closed circuit television network to 32 private and public high schools in Greater Manila, by META. The first year of operation (April 1, 1964 through May 30, 1965) "began with high hopes, sustained with enthusiasm, highlighted with a number of unexpected events and ended with optimism for the next operational year.

In 1964, the first series of training programs for studio teachers, classroom teachers, and school administrators of both

---

169. Dizard, op. cit., p. 246.

the Ateneo-Maryknoll closed circuit television network and META was offered; kinescoping of 112 half-hour lessons on physics began; and the first five telecasts for the CCTV Network was beamed to the Ateneo high schools on August 17, 1964 through channel 9 and channel 13. Preliminary appraisal and evaluation on technical operation, programming, pedagogical personnel and service indicated the effectiveness of telelessons. Problems in scheduling, television receivers requirements, and the quality of filmed telelessons were discussed and steps for improvements were taken.

Generally the benefits derived from instructional television were highly convincing to the administrators of the grade school and high school of both Ateneo de Manila and Maryknoll College.[170] The skeleton staff increased in 1965 almost three times over that in 1964 and requirements for additional outlets by both Ateneo high school and grade school were considered.

To test (1) the effectiveness of the televised courses in physics described above, as compared to conventional instruction, and (2) the acceptance of the project by the students, a limited research project was developed. The META, in order to carry out this project "employed the services of a research director who worked in consultation with the Ateneo Guidance Center. An achievement test, prepared by the studio teachers and two physics teachers, was administered to the televised (experimental) and conventional (control) groups at the end of the first and second semester."[171] A preliminary report after the first semester

---

170. "Ateneo de Manila University- The Philippines Center for Educational Television Annual Report", CETO News, (London) No. 11, June 1966, pp. 12-22.

171. Ibid.

indicated that as far as achievement was concerned, there were no significant differences between the experimental and control groups. Attitudes toward the televised courses were generally found favorable.

Television's educational activities during the year of 1964 and 1965 have enriched the experience of the Philippine education authorities. Educational administrators, teachers, students, and parents have been made aware of television as a new medium of instruction, and have in many ways indicated a growing interest in it.

According to the annual report, cited above, "the future of educational television in the Philippines seems practically certain now. There is no doubt that the use of educational television will grow and flourish and gradually expand to other cities and provinces of the country." Regarding the importance of the center, the report points out that "there is every indication that the influence of the center will make itself continually felt in this movement"[172] of growing of educational television.

---

172. Ibid.

CHAPTER FIVE

SUMMARY REVIEW OF USES OF EDUCATIONAL TELEVISION IN DEVELOPING COUNTRIES

I. INTRODUCTION

The new age of technology, coupled with the emergence, since World War II, of newly independent but underdeveloped countries, seeking a place in the sunlight of economic opportunity, has created a new and special urgency for improved education in these countries. The need is two fold--for the extension of educational facilities (for children and adults alike) to millions hitherto beyond its reach, and for the concurrent improvement of the quality and standards of education itself. According to one authority these needs are thus specifically described: (1) education must be planned to meet the economic needs and cultural aspirations of the people concerned; (2) education must be made available to all boys and girls, rural and urban dwellers, and must include plans for illiterate and poorly educated adults; (3) education must suit the planned development of the country, whether the needs are for students trained in technical subjects, home or rural economics, for the professionals or for other disciplines; (4) education should help people to think in global terms because they will be living an international life even though their home activity is based on a village background; and (5) education must be rooted in the social and cultural background of the country, as it must create a sense of the economic potential of the nation.[173]

---

173. "Educational Television in Developing Countries", op. cit., pp. 2-3.

The achievement of these goals, difficult even in the most advanced countries, is doubly difficult by virtue of the poverty, limited technical facilities (electricity, transportation, communications) and rudimentary educational resources that obtain, in varying degrees, in all the developing countries reviewed in earlier chapters. Common to them all are a shortage of skilled teachers, buildings, libraries, textbooks and other educational resources. It is the investigator's contention, and the hypothesis that this dissertation seeks to validate, that it is precisely these difficulties that invest the medium of television, and its proper use in formal and general education, with such promise and such varied potentialities for helping these developing countries win the race against time which all the nations of the world may be said to be running. In the following paragraphs we summarize the variant uses of educational television in developing countries that the investigator's researches have brought to light, and such evidence as is available.

II. VARIANT USES OF EDUCATIONAL TELEVISION AND THEIR INCIDENCE

The developing countries included in this investigation have apparently used educational television for one or more of the following purposes:

(1) In adult education: for (a) the teaching of illiterates and semi-literate adults as in Algeria and Kenya; (b) the teaching of fundamental education in health, agriculture and civic responsibilities as in India ; (c) teaching the development of cultural awareness of the national heritage and its development as in the evening programs in Uganda; and (d) disseminating information on other countries and world affairs as in the U.A.R.

(2) In instruction during school hours: including (a) team teaching as in India; (b) enrichment as in the Philippines; (c) direct teaching as in Argentina and Samoa; and (d) demonstration of scientific experiments not easily performed in conventional methods of teaching as in the Philippines. With regard to the grade levels television utilization has varied from country to country. In some countries it has been used from the elementary level to university such as in the Philippines, while in other countries, it has generally focused on elementary and secondary education as in American Samoa.

(3) In training provided in out-of-school hours as (a) helping teachers with lectures and demonstrations as practiced in the Philippines; and, (b) offering in-service training for teachers by showing well planned and well illustrated lessons, as well as with demonstrations of new techniques and methods by specialist-lecturers[174] as was experienced in Brazil, American Samoa, Nigeria, U.A.R. and India.

The following chart briefly illustrates the various uses of educational television:

A. FORMAL EDUCATION - School Instruction		
a. Kindergarten	-	Colombia, Argentina
b. Elementary	-	Brazil, Colombia, American Samoa, Nigeria, Uganda, Philippines.
c. Secondary	-	Colombia, Argentina, American Samoa, Nigeria, Ethiopia, U.A.R., India
d. College	-	Philippines, Argentina, Brazil
e. Teacher Training	-	American Samoa, Nigeria, Colombia, U.A.R., India, Brazil

174. Ibid., pp. 4-5.

B. INFORMAL EDUCATION - Adult Education		
a. Health, Hygiene,	-	U.A.R., India, Brazil, & General Agriculture Colombia, American Samoa, Uganda, Algeria, Liberia, Kenya, India
b. Illiteracy	-	Colombia, Brazil, Algeria, Kenya, U.A.R.
c. Nutrition	-	Colombia
d. Physical Education	-	American Samoa

Most of the countries surveyed have not been able to fully utilize television for educational purposes. Among the main problems they have been confronted with have been (l) signal limitations, i.e., the television transmittal range can cover effectively only an area of fifty miles from its originating points unless repeater station or stations linked with the main transmitter are used to carry the signal beyond the limited area; (2) electricity problem, i.e., television receivers, at least at the present stage, have to be linked to main supply of electricity, until prototype models of transistorized battery operated receivers become available; and, (3) the cost problem, i.e.., receivers are expensive and their installation is not justified except by a full integration of the television instruction into the planned educational pattern, and their continued operation safeguarded by a maintenance program.

As the preceding chapter indicates, there have been, generally, four principal ways in which education television has been used in developing countries. A system (l) where a television network is set up exclusively for educational purposes, as in American Samoa; (2) where commercial television stations carry educational programs as part of their public responsibility or through a contract with educational authorities, as in Colombia and Argentina; (3) where closed circuit television is set up by educational institutions for the purpose of education and instruction to be used exclusively by a particular institution or group of institutions such as at the University of Buenos Aires, or, its facilities are used for recording educational television for television transmission over commercial or national networks as in Manila in the Philippines; and, (4) where broadcasting has already been established as part of the governmental activities and educational television is a part of that commitment, as in India.

III. EFFECTIVENESS

The theoretical justification of television's use in education derives from its built in capacity to conquer space and thus to achieve two objectives never before possible in human history--communication to millions hitherto geographically space bound and therefore isolated, and the transmission of excellence (whether of knowledge and art, or of the teacher imparting such knowledge and such art) to the culturally privileged and unprivileged, the near and the distant alike. The specific application of television's use to the varied needs of those hungering for education have been summarized above. What remains is to consider and weigh the (educational) effectiveness of these uses of the medium.

It must be freely admitted that, on this side, the investigation has proved a disappointment. More evidence of television's viability as a tool of education had been anticipated than has in fact been forthcoming. Indeed, for those in search of "hard" evidence of a more or less scientific nature (evidence, for example, involving before and after measurements of knowledge acquired, comparative studies of experimental and control groups, etc.) the results of this study may seem in large measure fruitless, for evidence of this order is for the most part conspicuous by its absence.

To such appraisal of the worthwhile nature of this investigation the following considerations must be opposed. There is, first, the indisputable fact that education, truly conceived (i.e. conceived as more than mere rote learning), works its effects in ways too subtle to lend themselves to the precise measurement and tests dear to the scientist. Teachers, to this day, cast their bread upon the waters, hoping (only) that it will return to them after many days. Moreover, there is in education no instant measure of its effects, for some of these may show up only years later and in unforeseeable ways. Education, again, is more than the sum of all its parts and thus does not lend itself to enumeration of its effects by simple arithmetic. The elusiveness of education in this reference is perhaps best attested to by the fact that attempts to "measure" it have proved so inconclusive. In the United States, teaching by television has been extensively measured over against conventional teaching methods. For all that these tests have probed deeper than those normally used in classrooms, their overall findings have been that there is "no significant difference", as measured by results, between teaching by television (or, for that matter, a wide variety of other teaching techniques) and conventional teaching. Nevertheless, the use of instructional television has increased steadily in the United States, as in Japan and other technologically advanced countries, evidence in itself that other grounds have been found warranting its use. If these are held to be no more than subjective, presumptive or merely empirical, they are nevertheless (for the present, and until educational research advances far beyond its present competence) our operative guides in the determination of educational policy and practice. Some of these grounds, as related to this study's hypothesis, are cited below.

There is, to start with, the evidence that not one of the projects reported on has proved abortive. All have been continued and most expanded beyond their first, experimental stages. This is the more impressive as evidence in view of the heavy expenditures inherently involved in use of television.

With the cost factor again in mind, impressive is the number (see pages 33-38) of developing countries that either already use or aspire to using television.

Its viability is, third, attested to by the empirical judgment of those in charge of these many projects. Such testimonials as are cited in the individual reports on projects in Chapter Four, provide the evidence of men and women who have witnessed with their own eyes what has been achieved and further attest to how much more, given adequate resources of money and men, they believe can be achieved. It is the investigator's considered judgment that the accumulated evidence substantiates the hypothesis advanced in the first chapter of this dissertation and warrants the further inquiry, on which we now adventure, into which of the projects and procedures earlier reported apply pertinently to the needs and circumstances of Afghanistan.

CHAPTER SIX

CONDITIONS IN AFGHANISTAN AS COMPARED TO THOSE IN DEVELOPING COUNTRIES

This chapter will be divided into two parts: (1) an introduction dealing briefly with the history, the socio-economic and political conditions of Afghanistan; and (2) education, comprising the system, problems, and achievements .

I. INTRODUCTION

Afghanistan, a constitutional monarchy in the temperate zone of central Asia, touches the borders of Pushtunistan, Iran, Baluchistan, the People's Republic of China, and the U.S.S.R. in the east, west, south, northeast, and north respectively. Its borders comprise an area of 250,000 square miles, and its sparsely populated land is occupied by approximately 15,000,000 people.

The original population of the country is of Aryan stock,[175] the country being called Aryana in ancient history. Turks and Mongols, however, also live in northern and central Afghanistan, constituting a minority of the population as compared to the majority of Pathans. The two most widely spoken languages are Pushtu and Dari which belong to the family of Indo-European languages. The country is divided into twenty-nine provinces. The climate varies in some of the provinces to such an extent that school vacation starts in some parts in summer while in other parts in winter. With the exception of small minorities of Hindus and Jews,

---

175. Basic Facts on Afghanistan, American Friends of the Middle East, Inc., 1964, p. 5.

almost ninety-nine percent of the populations are followers of the Islam religion.

Agriculture, though primitive, is the main occupation. "In most of the country industrial technology at the end of World War II had not progressed beyond handicrafts and cottage industry, despite a high degree of skill in such industries as carpet weaving, leather wares, etc."[176] The textile plants, woolen plants, silk looms, hydroelectric power plants, margarine, soap and cement factories are producing but below the scale of production. The country is assumed to be rich in minerals but these are as yet unexploited.

The main exports include Karabul fur, dried and fresh fruit and nuts, raw cotton and wool, carpets and rugs, skin and hides, casings, and oil seeds, while the main imports are vehicles, most machinery, glassware, medical and chemical products, electrical appliances, and all rubber and petroleum products, all iron and steel products, as well as tea, spices and some sugar.[177]

The past history of Aryana, a name for ancient , Afghanistan, has been marked by a series of destructive invasions by various conquerors and subsequent reconstructions by its inhabitants, the details of which are irrelevant to the present study.

At the beginning of the twentieth century, however, a national awakening began and the first school for formal education was established. From 1919 to 1929 several educational institutions were started, progress in every aspect of life was noticeable, and people were motivated to work productively, students--male

---

176. Ibid., p. 21.

177. Ibid., p. 22.

and female--were sent abroad to acquire the skills necessary for further development and many progressive plansunder the leadership of His Majesty Amanullah were under way. This dynamic and progressive king, alas, lacked the skills of a good politician to control the situation while introducing new ideas to a tradition-bound culture.

The king's new plans, unfortunately, were looked upon with suspicion in certain circles and this attitude grew stronger and spread to the masses. Consequently the people rose against the king forcing him into exile. When, in 1929, the king had abdicated a brigand and his followers took Kabul. Turmoil and anarchy started, schools were closed and all the aspirations for progress were stifled. This anarchy would have continued if Nadir Shah had not rescued the country by overthrowing the regime, thus securing peace all over the country. To remedy the serious damages of the recent past the new regime founded by His Majesty Nadir Shah in 1931 had as one of its main tasks the reopening of schools, inaugurating of new ones, as well as motivating the people to participate in national development actively and optimistically. Thus, the history of organized education in Afghanistan starts in the 1930's.

II. EDUCATION

Educational institutions, from primary schools to the university level, are controlled and supervised by the centralized Ministry of Education whose headquarters are in Kabul, The head-masters, principals, and educational staffs in each of the twenty-nine provinces are directed and supervised by an agent of the Ministry. The schools are divided into three levels--primary (from grades one to six), secondary (from grades seven to nine), and Lycees or high schools (from grades ten to twelve). High school graduates are admitted to either of the two universities, Kabul and Nangarhar universities, on a selective basis. A number of vocational schools also exist for teacher training, nursing, agriculture, technology, trade and commerce, military training, theology and correspondence.[178]

Education is free at all levels. Primary education is compulsory as provided in the constitution.

> Education is the right of every Afghan and shall be provided free of charge by the State and the citizens of Afghanistan. The aim of the State in this sphere is to reach a stage where suitable facilities for education will be made available to all Afghans, in accordance with the provisions of the law. The government is obliged to prepare and implement a program for balanced and universal education in Afghanistan.
>
> It is the duty of the State to guide and supervise education. Primary education is compulsory for all children in areas where facilities for this purpose are provided by the State.[179]

Throughout the last decade, the Afghan government has been actively involved in pre-planning the economic development of the country in a series of Five Year Plans. The First Five Year Plan of Afghanistan started in September 1956 and ended in

178. Ibid., p. 18.

179. Constitution of Afghanistan, Kabul: Franklin Book Programs, Education Press, Article No. 34, October 1, 1964, p. 18.

1962, and the Second Five Year Plan started in March 1962 and will terminate in March 1967. Since educational development has been among the activities included in both plans, it will be appropriate to briefly survey the educational progress in the course of these two plans that encompass ten years from 1956 to 1967.

A. EDUCATION IN THE FIRST FIVE YEAR PLAN (SEPTEMBER 1956 - SEPTEMBER 1962)

A sum of Afs.[180] 10.30 billion, constituting 62 percent of government revenues and foreign loans and grants was expended on economic and social development for the entire First Five Year Plan, from which 7.7 percent was appropriated for education and health as compared to 49.5 percent for transportation and communication, 26.5 percent for industrial development, 12.6 percent for agriculture, and 3.8 percent for miscellaneous development works.[181]3 As a result of this planning "education made rapid strides, the number of students (primary, secondary, and vocational) rising from 96.34 to 169.06 per 10,000 of population. The number of students receiving higher education per 10,000 of population, rose from 0.66 to 1.44, and construction of a new campus for the Kabul University was taken in hand."[182]

The following table explains the educational progress of Afghanistan during the First Five Year Plan.

---

180. Afs. 50 is the equivalent of one U.S. dollar.

181. Survey of Progress, 1962-1964, Ministry of Planning, (Kabul: Franklin Book Programs, Education Press, August, 1964), p.3.

182. Second Five Year Plan (March 1962 - March 1967), Ministry of Planning, (Kabul: Government Printing House, 1962), pp. 1-2.

Number of Schools and Students 1956 and 1961				
	1956		1961	
Type of school	**Number of Schools**	**Number of students**	**Number of schools**	**Number of students**
Village				
boys	384	16, 421	693	39,463
girls	--	--	95	4,659
Primary				
boys	328	89,660	465	145,241
girls	20	10,012	68	23,735
Secondary				
boys	23	3,705	43	8,748
girls	4	881	9	2,431
Lycees				
boys	14	900	16	2,287
girls	2	188	6	628
Vocational				
boys	21	3,451	30	6,122
Kabul University				
boys	(7 depts.	830	(9 depts.	1751
girls	(	4	(	236
Total	804	126,092	1,434	235,301
				([183*])

The percentage increase in enrollment at the end of the First Five Year Plan was as follows:

Professional	education	6.9 percent
Secondary	education	16.3 percent

183. *Adapted from Survey of Progress, op. cit., pp. 147-151.

Higher	education	12.8 percent
Primary	education	7.0 percent
Rural	education	16.9 percent[184]

In spite of the above developments in education, only 7 percent of the working force had, by 1961, received primary education; while corresponding figures for vocational, secondary and higher education were 3.4, 1.0 and 0.5 percent respectively.

B. EDUCATION IN THE SECOND FIVE YEAR PLAN (MARCH 1962 - MARCH 1967)

Since the plan is still on its way, information pertinent to education or any other aspects of economic development has not yet been made available. The Ministry of Planning, however, has issued a booklet that covers the first two years of the Second Five Year Plan, i.e. from 1962 to 1964. The total outlay for the plan has been estimated at 44,500,000 Afghanis which is more than three times the expenditure during the First Five Year Plan. Total outlay amounts to 31,353 million Afs. representing about a four-fold increase over investment in the first Plan. The investment for education has been estimated at Afs. 2,190,000,000, which is roughly 8 percent of the total investment.[185]

The aims of the Ministry of Education in the Second Five Year Plan have been described as encompassing (1) the spread of literacy throughout the country by increasing the number of elementary schools; (2) the encouragement and popularization of advanced scientific, technical and professional education; and (3) the training of personnel required for implementation of devel-

---

184. "Second Five Year Plan", op. cit., p. 56.

185. Ibid., pp. 10-11.

opment plans.[186] These aims are similar to those chosen for the First Five Year Plan.

In the first two years of the Plan (1962-1964) considerable progress in the field of education has been reported. 172 village schools, 35 elementary schools, 34 new secondary and high schools were established throughout the country and the number of students in these schools was increased by 21,000, 42,000, and 5,700 respectively. In addition to this, the establishment of 22 new professional schools with a total enrolment of 6,400 students was also reported.[187] At the university level an increase of 385 students in social studies and 131 students in science classes has been reported. Graduates of Kabul University numbered 305 in 1962; this figure represents a 3 percent increase in the total number of graduates as compared to the year 1961. The number of college graduates reached 399 in 1965.

In spite of the authorities' serious concern with respect to education, educational efforts have always fallen short of educational demands and needs in Afghanistan. This ever-increasing demand for more and better education in an ever-increasing: population has been so unpredictably pressing that, as mentioned earlier, it was virtually impossible for the experts of a combined committee of the United Nations Scientific and Cultural Organization (UNESCO) and the Economic Commission for Asia and Far East (ECAFE) to predict figures for enrollment sufficiently accurate to permit adequate future planning. The committee predicted in 1963 that the number of elementary school students in

186. Ibid., p. 67.

187. "Survey of Progress", op. cit., pp. 66-67.

1965 would be from 275,000 to 303,000. By 1964 this prediction had been exceeded considerably.[188]

Afghanistan's general attitude toward education and modernization is perhaps more propitious for future development than that of some tradition-bound countries in the developing regions. The people's demand for more and improved education and their cooperation with the educational institutions are part of the evidence reflecting the nation's desire to adopt new methods, techniques and ideas in their educational system.

Socially, vocational opportunities are widest for those equipped with the best education. The educated person is likely to secure a higher salary and face a brighter future both in the civil service and in industry. Culturally, educated individuals are more highly regarded in terms of their cultural and traditional status in Afghanistan. The intellectual has been respected and serves as an ideal for the young. Religiously, Islam, the dominant religion in the country, greatly emphasizes education in its teaching and regards it as compulsory for both sexes. These factors, among others, consequently led to increasing requests for education. The educational authorities are forced to meet this pressing educational demand. They try to explore the most effective means to correct the situation and to mitigate the following shortcomings:

1. Teachers

In spite of attempts to hire part-time teachers, to employ foreign experts, and even to use the talents of immature graduates of elementary and secondary schools, the shortage of teach-

---

188. Aref Ghousi, translator, "The Role of Education in Economic Development," Irfan, (Kabul), No. 9, (January 1966), p. 27.

ers has been one of the serious problems of Afghanistan. The supply of qualified teachers is even more limited. Although the Columbia Team, a group of experts from Teachers College, has been helping the Ministry of Education to train qualified teachers since 1949, the shortage of teachers has continued unabated.[189] Among the main factors that keep the number of teachers limited are perhaps that (1) teachers are poorly rewarded and some of the active teachers, in order to balance their expenditures, have part-time jobs outside their respective schools; (2) educated teachers with training in education outside the country are either posted or they themselves try to be posted to other civil service departments where, if money does not make the difference, prestige does; and (3) teacher training institutions are not producing enough graduates to cope with ever-increasing demands on teaching personnel.

The existing teacher training institutions are reported to have produced altogether 850 teachers in 1965 to teach in 2086 villages, elementary, secondary, high, and vocational schools throughout the country.[190] The shortage of teachers compelled the Ministry of Education in 1965 to ask the teachers of different provinces to volunteer in offering their services during the three months of their vacation. According to this plan, teachers who teach in a hot climate and receive their vacation in summer-time would teach in provinces where the schools are on a vacation in winter-time and vice versa. Although this exchange of volunteers is but for a limited period of three months, according to the Ministry of Education, it is better than having classes without teachers

---

189. Irfan, op. cit., No. 9, November, 1965, p. 2.

190. Irfan, op. cit., No. 7, August, 1965, p. 65.

of any kind, or having unqualified teachers who need to be helped by the volunteers.[191]

In spite of these efforts the shortage of teachers has been such that in some schools the service of sixth grade graduates as teachers has been resorted to. Indeed, as late as 1965, there were elementary teachers in the provinces without any certificate or formal schooling of any kind. It is this fact which prompted a resolution at the annual meeting of teachers in the Ministry of Education.

> The official appointment of sixth grade graduates or those who have been considered the equivalent of sixth grade graduates without having formal schooling, should be discontinued. However, if the necessity is severe, the appointment should be based on a contract basis. Nevertheless girl schools of the provinces should be excepted from this rule.[192]

2. Buildings

In spite of the fact that in most parts of the country the people volunteer their cooperation by donating lands and money for school buildings, the educational authorities, due to their limited budget, have not been able to build sufficient school-buildings. Most of the schools, particularly elementary and secondary schools, are without school buildings of any kind, and the stu-

---

191. Irfan, op. cit., No. 9, November, 1965, p. 3.

192. "The Proposals of the Provincial Teachers to the Ministry of Education, During Their Annual Meeting in Kabul in 1965", Irfan, op. cit., No. 7, August 1965, Franklin Printing Programs, Education Press, p. 8.

dents are taught at different hours on a shift system. Many are almost totally destitute of modern facilities and lack proper lighting or adequate classrooms.[193] The lack of school buildings is such that in the provinces of Afghanistan the mosques have been used as school-buildings.

> The Mosque in Islamic religion is not only a place of worship, but also the center of fundamental education for children as well as adults. This institution is fully supported and maintained by the community. Every village has a mosque. The mosque is a very important source of knowledge and education if it is properly rated and utilized. My Ministry is using these facilities for schooling and it can be used as center for adult education, and literacy classes with little or no expense. Considering the number of mosques available and the support received by these from the nation, I think it should be a dynamic force if properly handled to serve the cause of eradicating illiteracy.[194]

3. Libraries, Textbooks, and Other Educational Facilities

There is not a single well organized library with sufficiently up-to-date material for an elementary, secondary, or high school system in Afghanistan. There are, however, a limited number of libraries containing materials which are either old or printed in languages that cannot be used by the local population. There has

---

193. "United Nations Education Mission", op. cit., p. 20.

194. Mohammad Anas, Royal Afghan Minister of Education, A speech at the opening session of the World Congress of Ministers of Education, held in Teheran, September 8-19, Irfan, op. cit., No. 7, August 1965, p. 83.

never been an adequate number of textbooks available. Paper is expensive and printing costs are so high that publishers are unwilling to risk producing them. School and even college students, therefore, depend solely on notes dictated by their teachers. Although the South East Asian countries, where one quarter of the world population lives, produce, only one-fortieth of the total books published in the world, Afghanistan has the lowest production among the countries of South East Asia.

Country	**Annual Publishing of Books**
Afghanistan	60
Burma	603
Cambodia	230
Ceylon	1,969
Indonesia	869
Iran	569
Malaya	338
Philippines	153
Singapore	237
Thailand	1,397
Viet Nam (Republic)	1,551[195*]

Education suffering from these inadequacies and inequities has no relation to life and therefore fails to affect, in any significant manner, the development of the country. Although the First and the Second Five Year Plans have had their influence in the betterment of education, the serious problems have not been satisfactorily alleviated. The present educational problems and particularly the more tense emerging problems of the future,

195. "Books for the Developing Countries of Asia", Reports and Papers on Mass Communication, No. 47, UNESCO, (Paris), 1965, p. 5.

make the prospect of educational activities bleak, if modern technology is not employed.

CHAPTER SEVEN

CONCLUSIONS AND RECOMMENDATIONS

Widely dispersed, as they are, across the world, and diverse as are their respective histories, their cultures and their social and political organization, the countries surveyed in this investigation nevertheless exhibit certain striking similarities. These both distinguish them from more advantaged countries and determine, for each and all, their dominant needs and aspirations. These similarities are so starkly evident as neither to need nor to warrant elaborate substantiation beyond the facts earlier brought to light. All these countries are marked by (1) the wide incidence of extreme poverty; (2) backwardness in the use of, and shortage of the equipment and know-how of modern technology; (3) desperate deficiencies in education with respect to all its fundamental components--buildings, text-books, teachers, etc.

> ...Education in our country today faces the same problem that many school systems all over the world are trying to solve--the steadily growing number of persons who must be educated, the tremendous pace at which new knowledge must be acquired, the lack of qualified teachers, the lack of school buildings and classroom equipment.[196]

Differences, in these matters, between these countries are only a question of the degree of misery involved--as it is also when we compare them with Afghanistan. In what follows, therefore, an attempt has been made to formulate an answer to the follow-

---

196. Vitaliano Bernardino, op. cit., p. 147.

ing questions: (1) Should Afghanistan make use of television for its educational activities? and if it should then (2) Which of the methods of utilizing educational television adopted in developing countries would appear to be the most suitable pattern for educational television in Afghanistan?

A. SHOULD AFGHANISTAN MAKE USE OF TELEVISION FOR ITS EDUCATIONAL ACTIVITIES?

The investigator believes that the conditions in and the needs of Afghanistan, being substantially the same as in the developing countries surveyed, television can bring about a crucial change in Afghanistan's educational structure. The present educational problems in Afghanistan, as explained earlier, are so tense that the future of education seems dreadful unless the technology of television is applied. Television under the present circumstances seems to be a viable contributory answer to the problems that handicap education, and consequently hold back the country.

B. WHICH OF THE METHODS OF UTILIZING EDUCATIONAL TELEVISION ADOPTED IN DEVELOPING COUNTRIES WOULD APPEAR TO BE THE MOST SUITABLE PATTERN FOR EDUCATIONAL TELEVISION IN AFGHANISTAN?

This particular portion of the study involves recommendations for the possible methods of utilizing educational television in Afghanistan. These recommendations draw on the facts already explored. Prior to making such recommendations re-statement of the alternatives and exploration of their possible adoption for Afghanistan seems necessary. In this respect the questions of what kind of system, what subject matter and service to what audience will be discussed and possible action will be recommended.

1. The System

There have been, generally, four principal systems adopted for educational television in developing countries. Afghanistan has already adopted the fourth system, a system that brings broadcasting strictly under government ownership and control. This system is similar to the Indian broadcasting system, in incorporating all broadcasting activities as a public enterprise directly under the control of the Ministry of Culture. Radio Afghanistan has already been established under this system and similar provisions have been envisaged for television.

Since a closed circuit television system will confine the reception of televised instruction to a comparatively limited number of school and college students (except as phenomenal sums of money are expended), this system in Afghanistan, where adult education is equally as important as school instruction, does not seem to be advisable. Afghanistan, a country with one of the highest rates of illiteracy in the world, and with a very limited number of publications in circulation, needs open-circuit broadcasting to reach as wide an audience as possible. This system of open-circuit broadcasting seems to be, in the long run, considerably less expensive and more efficient than closed-circuit one.

Since broadcasting activities are under the Ministry of Culture, while education is controlled by a centralized Ministry of Education in Afghanistan, it appears appropriate that televised instruction and education be a combined effort of the two Ministries, as is the case between broadcasters and educators in the Philippines. Educational programs should be prepared and planned by the Ministry of Education, and production, technical facilities as well as broadcasting should be the responsibility of the Ministry of Culture.

Seeing that access to a receiver is the condition of service by television, and considering the abject poverty of most people in Afghanistan, the government should provide receivers for sale on easy terms or rent for those who cannot afford to buy a set. This system of providing television sets on rental basis has been quite a satisfactory system in Liberia. Similarly the government should envisage provision of sufficient numbers of receivers for public viewing in restaurants, tea-houses and cafes as was done in Italy and Japan. In addition to these provisions, the government has to provide receivers free of cost to be installed in schools, auditoriums, cinemas, public parks, and other recreational areas.

2. The Subject Matter

Choosing subject matters for educational television requires thorough research to determine what subjects should be given priority, what should be the target area, and what subjects best lend themselves to the medium. Judging from the educational problems in Afghanistan described above, the investigator believes that educational television in Afghanistan should serve two primary goals: broadcasting for schools in and out of school hours, and broadcasting for adults. Since the most pressing need for education in Afghanistan--due to the highest rate of illiteracy and the great need for adult education pertaining primarily to such subjects as health, hygiene and agriculture--is in elementary education, educational television should start on the elementary level. Telecasting of educational material on the elementary level is more advantageous, as compared to broadcasting on higher levels, since its contents can easily be understood by adults as well, as was effectively demonstrated in Algeria. To handle the subject matter efficiently, teacher training centers and workshops, as was done in the United Arab Republic, should be organized. Teacher training by television has been practiced in Brazil, Samoa, Nigeria, Colombia and India as well.

3. Possible Ways of Handling Initial Problems

Educational television, as the investigator believes, should start on a small scale and expand gradually when the time and economy permit. The first station should be installed in Kabul, the capital of the country, for the following reasons:

(a) Kabul is the most densely populated city in the country and televised instruction and education can be received there by a greater number of viewers than in any other city.

(b) Kabul has more educational institutions than any single city in Afghanistan. In addition to this, Kabul University is also located in Kabul.

(c) The radio headquarters are all located in Kabul, and creation of educational television there facilitates the exchange of technicians between radio and television organizations.

(d) The Ministry of Education is in Kabul, and this situation eases the problem of planning and preparing educational material for telecasting.

(e) The City of Kabul itself is dominated by two high hills that make excellent natural broadcasting towers, thus, simplifying coverage problems.

(f) The city can be covered by a medium power station. A station capable of covering a radius of 20 miles, can cover the whole city plus some of the immediate suburbs in its transmittal coverage.

(g) Most of the suburbs of Kabul have electricity, and by kinescope or tape recording the televised instruction or education can become storable and reusable on television stations as well as they can be shown through 16 mm projectors even to those communities that are beyond the reach of the television spectrum.

APPENDIX

Source: Research Report of the United States Information Agency, Research and Reference Service, June, 1966

- 182 -

World's TV Sets Growth Since 1955
(U.S. & Canada Excluded)

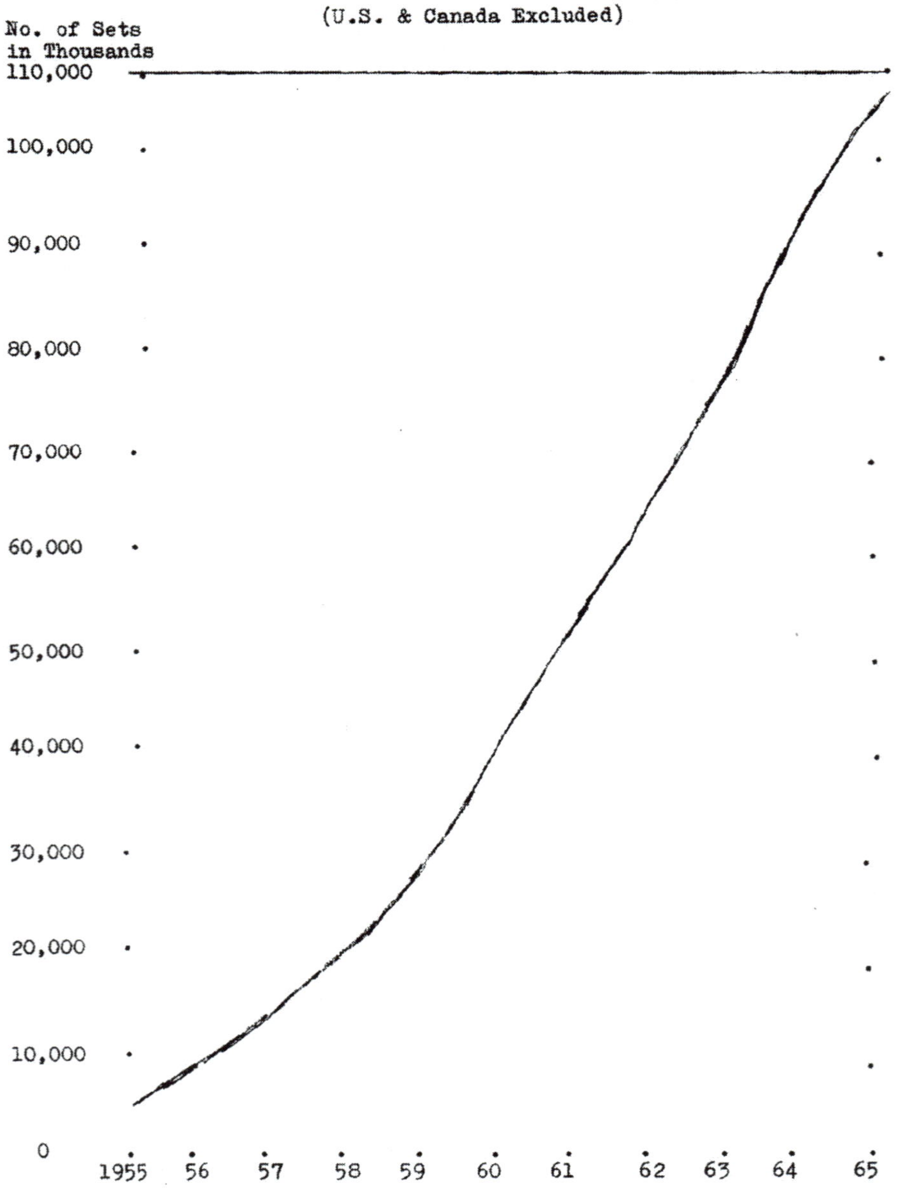

Sets in Thousands

World's sets Growth since 1955, by areas

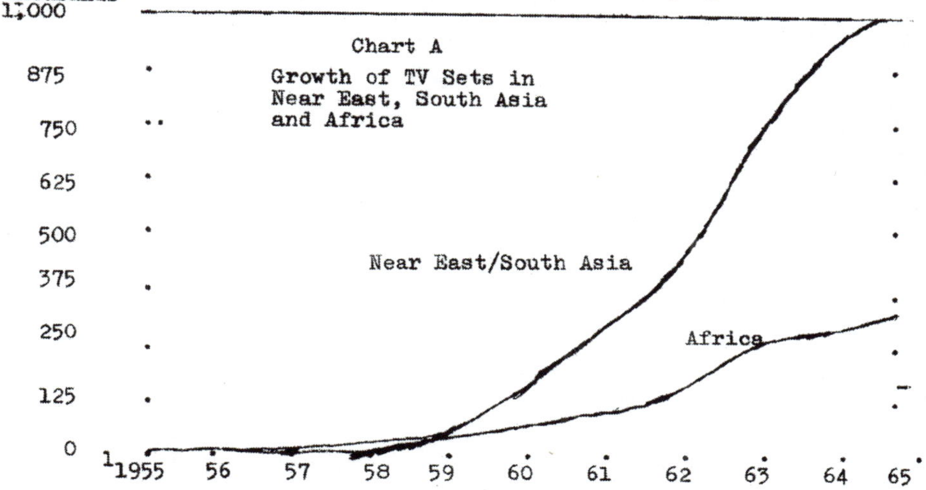

Note difference of scales between char A & B

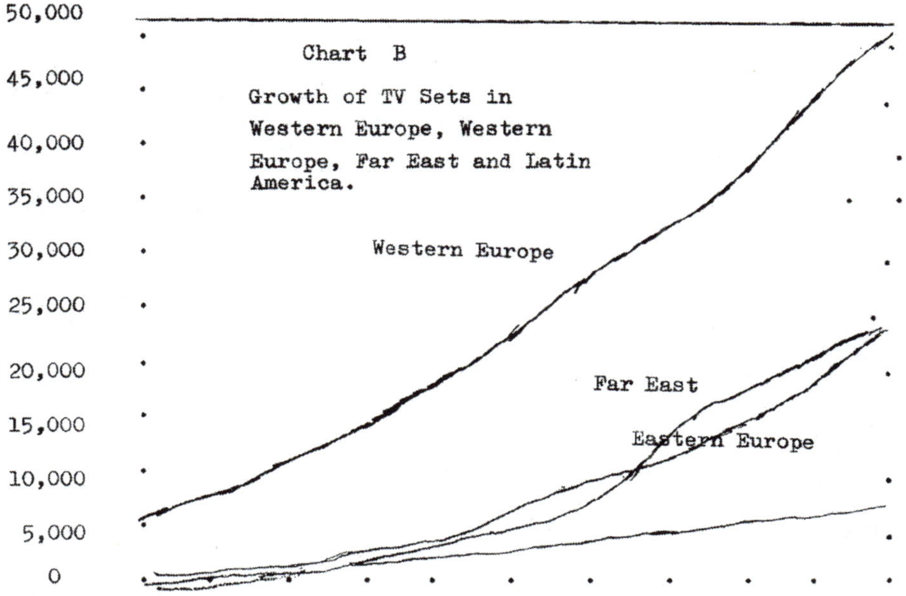

Growth of Overseas Television by Areas - 1955 - 1965 Television Receivers

Date	Western Europe	Eastern Europe	Near East & South Asia	Africa
12/55	6,018,400	1,063,200	200	5,000
12/56	8,364,100	1,481,800	800	5,000
12/57	11,341,200	2,349,000	4,100	11,000
12/58	14,676,500	3,321,000	11,200	25,000
12/59	19,053,900	5,303,900	45,800	43,800
12/60	23,816,800	7,404,600	150,900	69,300
12/61	29,189,000	9,406,000	306,000	96,000
12/62	33,581,700	11,404,000	408,400	128,300
12/63	39,033,200	15,283,400	724,800	250,100
12/64	45,931,600	19,704,000	938,800	277,100
12/65	50,942,700	23,581,400	1,039,200	313,300

Date	Far East	Latin America Caribbean	GRAND TOTAL
12/55	259,700	619,000	7,965,500
12/56	487,600	1,190,000	11,529,300
12/57	1,124,000	1,560,800	16,390,100
12/58	2,530,500	2,314,500	22,878,700
12/59	5,118,000	2,524,600	32,090,000
12/60	7,946,200	3,553,600	42,941,400
12/61	10,241,000	4,522,000	53,760,000
12/62	14,796,800	5,182,700	65,501,900
12/63	18,894,700	6,142,800	80,329,000
12/64	20,977,200	6,645,700	94,474,400
12/65	23,842,300	7,548,200	107,267,100

World's Television
(1965)

Areas	Receivers in Use	
	1/1/65	12/31/65
Western Europe	45,931,600	50,942,700
Eastern Europe	19,704,000	23,581,400
Near East & South Asia	938,800	1,039,200
Africa	277,100	313,300
Far East	20,977.200	23,842,300
Latin America & Caribbean	6,645,700	7,548,200
TOTAL	**94,474,400**	**107,267,100**

World's Television Receivers
by Countries (1965)

WESTERN EUROPE	
Austria	728,600
Belgium	1,461,300
Denmark	1,079,300
Finland	731,900
France	6,517,500
West Germany	11,379,000
Gibraltar	5,200
Iceland	10,000
Ireland	370,000
Italy	6,216,500
Luxembourg	31,000
Malta	29,500

Monaco	15,000
Netherland	2,239,100
Norway	490,000
Portugal	181,800
Spain	1,750,000
Sweden	2,084,900
Switzerland	622,100
United Kingdom	15,000,000
TOTAL	**50,942,700**

World's Television Receivers by Countries (continued)

EASTERN EUROPE	
Albania	1,000
Bulgaria	187,800
Czechoslovakia	2,240,000
Germany (East)	3,216,400
Hungary	831,200
Poland	2,077,800
Rumania	450,000
USSR	14,000,000
Yugoslavia	577,200
TOTAL	**23,581,400**

NEAR EAST AND SOUTH ASIA	
Aden	10,000
Bahrein	7,000
Cyprus	14,400
Greece	3,500
India	1,000
Iran	110,000

Iraq	170,000
Israel	15,000
Jordan	3,500
Kuwait	45,300
Lebanon	136,000
Pakistan	1,800
Qatar	2,500
Saudi Arabia	30,000
Syrian Arab Republic	66,000
Turkey	1,200
United Arab Republic	422,000
TOTAL	**1,039,200**

AFRICA	
Algeria	150,000
Congo (Brazzaville)	400
Congo (Leopoldville)	500
Ethiopia	4,200
Gabon	400
Ghana	800
Ivory Coast	2,000
Kenya	9,900
Liberia	2,500
Libya	5,000
Mauritius	4,000
Morocco	31,000
Nigeria	30,000
Reunion	2,000
Senegal	100

Sierra Leone	1,100
Southern Rhodesia ..	39,000
Sudan	10,000
Tunisia	5,400
Uganda	4,700
Upper Volta	100
Zambia	10,200
TOTAL	**313,300**

FAR EAST	
Australia	2,457,300
Communist China	100,000
Hong Kong	49,000
Indonesia	45,200
Japan	20,000,000
Korea (South)	50,600
Malaysia (including Singapore)	114,300
New Zealand	412,900
Philippines	180,000
Ryukyu Islands (Okinawa)	120,000
Taiwan	63,000
Thailand	250,000
TOTAL	**23,842,300**

LATIN AMERICA & CARIBBE-AN	
Argentina	1,500,000
Barbados	4,500
Bermuda	14,000
Brazil	2,500,000
Chile	50,000
Colombia	400,000
Costa Rica	40,000
Cuba	500,000
Dominican Republic	52,500
Ecuador	34,700
El Salvador	30,000
French Antilles	6,300
Guatemala	46,200
Haiti	5,500
Honduras	8,000
Jamaica	22,100
Mexico	1,215,100
Netherlands Antilles	25,000
Nicaragua	10,000
Panama	75,000
Paraguay	3,100
Peru	220,000
Surinam	5,600
Trinidad and Tobago	20,100
Uruguay	185,000
Venezuela	550,000
TOTAL	**7,548,200**

BIBLIOGRAPHY

I. BOOKS AND PERIODICALS:

1. Apple, R. W. "What Telstar Cannot Do?", Saturday Review, (September 8, 1962)
2. "Ateneo de Manila University-The Philippines Center for Educational Television Annual Report", CETO News, (London), No. 11, June, 1966, pp. 12-22.
3. Boyer, Sir Richard, "TV in Perspective," Australian Quarterly, (September, 1957)
4. Bernardino, Vitaliano, "Director of Public Schools Department of Education, "A Brief History of the Development of Instructional Television in the Philippines", Tokyo, 1964.
5. Cassirer, Henry, "TV Crossroads," Television Quarterly, Summer 1963.
6. ---- Television Teaching Today. Paris: UNESCO, 1960.
7. Chander, Ramesh, "Sound and Television in the Fundamental Instruction of Adults-Citizenship through Television, All India Radio", Proceedings of the Second International Conference of Broadcasting Organizations on Sound and Television, Tokyo, 1964.
8. Chen Shih-ping. "TV Comes to Taiwan," Free China Review, (November, 1962)
9. Clarke, Arthur C. Extra Terrestrial Relays: Can Rocket Stations Give World-Wide Coverage?" Wireless World, (London: October, 1945).

10. Cooding, George A. "Broadcasting Without Barriers," (Paris: UNESCO, 1959)

11. Diamond, Leslie A. W. "Bringing Radio and Television to Northern Nigeria," EBU Review (Geneva: No. 93 B, September, 1965)

12. Dieuzeide, Henri. "TV Education in France," Realities, (July, 1963)

13. ---- "Ten Years of School Television," EBU Review, (Geneva: No. 69B, September, 1961).

14. Dizard, Wilson P. The Strategy of Truth, Washington: Public Affairs Press, 1961.

15. ---- Television: A World Review. New York: Syracuse University Press, 1966.

16. Doherty, D. M. "TV Italiana," Sponsor, March 28, 1959.

17. ---- "TV in Japan," Sponsor, June 6, 1959.

18. ---- "TV in Thailand," Sponsor, June 20, 1959.

19. Duhoura, S. J. Carlos A. "Technical School Television in Argentina," Television and Adult Education (Paris: No. 13, March, 1964).

20. Dunlap, Orrin E. Jr. Outlook for Television, New York: Harper & Bros., 1932.

21. Eastern Economist (New Delhi). "The Challenge of Television, "Special Supplement, Nov., 1959.

22. Educational Television in Developing Countries, A Report of An On-the-Spot Survey by Australian Broadcasting Commission and Nippon Hoso Kyokai, 1964.

23. EBU Review (Geneva) "Aid to the Broadcasting Organizations of the Developing Countries," No. 88 B, November, 1964.

24. EBU Review (Geneva) "Origin and First Steps of the EBU Program Committee," No. 85 B, May, 1964.

25. EBU Review (Geneva),No. 91 B, May, 1965.

26. Far Eastern Economics Review (Hong Kong) "Television for Asia," Special Supplement, January 16, 1964.

27. Faust, Clarence H. "Educational Philosophy and Television," The Educational Records, vol. 39, No, 1, January, 1958.

28. Fernig, L. R. and McDougall, J. F. "Trends in World Education," Phi Delta Kappa, vol. 37, No. 9, June, 1958.

29. Finocchiaro, Ing. Jose, Director of Telescuela Education and Television in the Republic of Argentina. CETO News (London), No. 10, (March, 1966).

30. Ghousi, Aref, "The Role of Education in Development of Economy", Irfan; Kabul No. 9, January 1966.

31. Givton, Hanock. "Introducing Television into Israel," EBU Review (Geneva), No. 93 B, September, 1961

32. Grant, Douglas. "Schools Television in Western Nigeria," EBU Review (Geneva), No. 69 B, September, 1961.

33. Greenhill, L. P., "Learning Resources for Higher Education" (London), Medical and Biological Illustration, October, 1964. Vol. XIV, No. 4.

34. Hadsell, R. S. and G. K. Smith, "Educational Television in Iran, "Multiplier, U.S. International Cooperation Administration, 1961.

35. Hansen, George, "Eurovision", Telefilm International (Los Angeles), January, 1960.

36. Halstead, William S. "Low Cost National Television Comes to East Africa," Multiplier U. S. International Cooperation Administration, June, 1964.

37. Hunt, Harold C., assisted by Steward David. Television and Formal Education, in William Y. Elliott (ed.), Television Impact on American Culture. East Lansing, Michigan State University Press, 1956.

38. "In India", Television and Adult Education, (Paris), No. 13, May 1964, pp. 28-29.

39. Johnsen, O. Lund (ed.). World Radio-TV Handbook. Eighteenth edition, Hellerup, Denmark, 1964.

40. Ealimullah, A. F. "The Future of Television in Pakistan," CETO Review, No. 6 London: Center for Educational Television Overseas, March, 1965.

41. Kurzwell, Z. E. "Television in Israel," Television and Adult Education (Paris) No. 12, February, 1964.

42. Lambert, R. S. "The Challenge of Television in Education," Canadian Education, vol. 13 No. 3, June, 1958.

43. Laslett, Peter. "Learning from Television", Twentieth Century (London) Autumn, 1964.

44. Lerner, Daniel. The Passing of Traditional Society, Glenco, Ill.: Free Press, 1958.

45. Linz, C. C. "Television and Education", Australian Quarterly, September, 1958.

46. Mallikan, Max and Donald L. M. Blackmer. The Emerging Nations. Boston: Little, Brown and Co., 1961.

47. Martinez, A. M. "TV in Latin America," Advertizing Agency, October, 1951

48. Mathur, J. C. "First Steps Towards Educational Television, "EBU Review (Geneva) No. 69 B September, 1961.

49. Mills, J. G., TV Education Officer, "Delhi ELT/TV Project", CETO News, (London), No. 7, June, 1965.

50. Menon, V. K. Naryana, Director General of All India Radio, "India in the Approach of a Secular State, EBU Review, (Geneva), 97 B, May, 1966.

51. Moir, Guthrie and John Lord. "Teaching by Television," New Commonwealth, September, 1959.

52. Nassr, Bahir. "Television in Egypt," Broadcasting, June 10, 1963.

53. Patridge, E. Dealton. Continental Classroom: (One Answer to a National Need," Audiovisual Instruction, Vol. 4, No. 3, March, 1959.

54. Powell, John Walker, Channels of Learning, (Washington D. C.: Public Affairs Press, 1962).

55. Peace Corps, The Office of Radio and Television Programmes, Colombia ETV, a report, Washington, D. C, January, 1966.

56. Peace Corps, Colombia Educational Television, Program Description, Peace Corps, Washington, D. C., 1965.

57. Puglisi, Maria Grazia. "Telescula Enters Its Second Year," EBU Review (Geneva),No. 59 B January, 1960.

58. ---- "Television and the Fight Against Illiteracy, "EBU Review (Geneva) No. 82 B November, 1963.

59. Queenan, B. P., Senior Producer, CETO, "Development in Educational Television in Latin America, CETO News, No. 11, June, 1966.

60. Ramasastri, J.V.S., "Nationalization of Broadcasting in India," Indian Economic Journal, July, 1959.

61. Rengelink, J. W. "The European News Transmission - A First Step in World-Wide News," EBU Review (Geneva) No. 91 B May, 1965.

62. Rao, K. N, and Wickham, Robert S., A Review of Educational Television in Latin America, report to the Ford Foundation, November 28, 1966.

63. Romanelli, Ricardo A., Closed Circuit Television in the University of Buenos Aires, E.T.V. Department, CETO News (London), No. 10, March, 1966.

64. Rosen, George. "TV - The Best of Goucha," Variety, October 10, 1962.

65. ---- "Master Rides Again," Variety, January 13, 1965.

66. Ross, Gordon. Television Jubilee: The Story of 25 Years of BBC Television. London: W. H. Allen, 1961.

67. Schramm, Wilbur. Mass Media and National Development. Stanford, California: Stanford University Press, 1964.

68. ---- "What We Know About Learning from Instructional Television, Educational Television; The Next Ten Years, (Stanford, California: Institute for Communication Research, (California: Stanford University Press, 1962).

69. Siepmann, Charles A., Television and Our School Crisis, (New York: Dodd Mead and Company, 1958)

70. Shayon, Robert Lewis (ed.). The Eight Art New York: Holt, Rinehart and Winston, 1962.

71. Second Five Year Plan (March 1962 - March 1967), Ministry of Planning, (Kabul: Government Printing House, 1962).

72. Souza, Alfredina de Paiva e, professor at the Institute of Education, Rio de Janeiro, "Educational Television: A Brazilian Experiment", Television and Adult Education. (Paris, 1965).

73. Survey of Progress, 1962-1964, Ministry of Planning, (Kabul: Franklin Book Programs, Education Press, August, 1964).

74. Television Age. "Television in Orbit," A special issue devoted to international television, July 19, 1965.

75. "Television in India", Indian and Foreign Review, Vol. 2, No. 18, July 1, 1965.

76. Television Information Office. Television and Education: A Bibliography. New York: December, 1960. Pamphlet.

77. Tuber, Richard, "A Survey of Programming on the Central Studios of TV, Moscow, January-June, 1960", Journal of Broadcasting, (Fall, 1960).

78. UNESCO, Mass Communication (press, film, radio, television) 1965.

79. UNESCO, Education, publications, 1965.

80. UNESCO, reports and papers on Mass Communication, 1965, as published in "Current List UNESCO publications, 1965".

81. United Kingdom Colonial Office. Sound and Television Broadcasting in the Overseas Territories. 1964 Handbook. 16th edition London: Information Department, Colonial Office, 1964.

82. United States Information Agency. Overseas Television Developments. Annual Review of Global Television Development issued each April since 1959, Washington, D. C.

83. USSR in Figures, (Moscow: Central Statistical Publishing House, April, 1965).

84. Van Dalen, D. B., Understanding Educational Research, (New York: McGraw Hill Book Co., Inc. 1962).

85. Varnum, Walter C. "ETV: Messiah or Monster?" The American Teacher Magazine, October, 1958

86. Watson, J. "United Nations TV," Television Magazine, October, 1958.

87. Whitehead, Alfred North, The Aims of Education, (New York: Macmillan Company, 1929).

88. Williams, F. N. Lloyd, "School Broadcasting, Sound", EBU Review (Geneva), No. 70 B (November, 1961).

89. Wagley, Charles, An Introduction to Brazil, (New York: Columbia University Press, 1963).

90. World Communication: Press, Radio, Television Film. Paris: UNESCO, 1964 edition.

91. Yench, John. "Role of Radio and Television in Philippines, American Chamber of Commerce Journal, Manila, December, 1960.

92. Zender, Harry. "Television in Israel," Gazette (Amsterdam) No. 1, 1961.

II. CONFERENCES, SEMINARS, HEARINGS

93. Ahmad, Abbas. The Part Played by Television in Developing Countries. A paper submitted by the United Arab Republic Delegation to the third International Television Festival, 1964.

94. Dale Edgar, Problems in Introducing New Media into Education, in: Conference on ETV. Washington, D. C.1958. Proceedings.

95. Educational Television: The Story Up to Now, Speech Before the Second Annual Radio-TV Conference, Boulder, Colo., April 7, 1959.

96. National Association of Educational Broadcasting. Seminar on instructional uses of television and radio. Purdue University, July 21-25, 1958. Edited by Elaine Afton, 1958.

97. Television and Adult Education (Paris). Meeting on the Introduction and Development of Television in Africa. No. 15, October, 1964. Report of a UNESCO seminar held in Lagos, Nigeria. September, 1964.

III. REPORTS, SURVEYS. AND ARTICLES

98. Barnell, M. "TV Spreading East Through Central America," Barrons', June 4, 1956.

99. British Broadcasting Corporation, BBC Educational Broadcasting. Outlook Educational Programmes for Adult Listeners and Viewers, London, 1963.

100. ---- BBC Educational Broadcasting. Report to the Delegates of Second World Conference on School Broadcasting. Tokyo. London: April, 1964.

101. Dieuzeide, Henri. Teaching Through Television Paris : Organization for European Cooperation, Office for Scientific, Technical Personnel, 1960.

102. Dunlap, A. D. The Development of Television in the Middle East, Africa, and South Asia. Cambridge: Center for International Studies, Massachusetts, Institute of Technology, April, 1957. Mimeo.

103. Ford Foundation. ETV: A Ford Foundation Pictorial Report. March, 1961.

104. Huth, Arno. Communications Media in Tropical Africa. Washington: United States International Cooperation Administration, 1961 Government financed study of media in 14 African countries.

105. Hurley, S. J. Neil. "Airborne Television and Universal Primary Education," Santiago, Chile: Centro de investigation y accion social, 1964. Mimeo.

106. Furu, Takeo. Television and Children's Life - A Before After Study. Tokyo: Radio and Television Culture Research Institute, Japan Broadcasting Corporation, 1962.

107. Larkin, S. J., Leo H. Educational Television for the Ateneo de Manila: A Preliminary Study. New York: Fordham University, May, 1960. Mimeo.

108. ---- A Report on ETV for the Philippines. Manila: Ateneo de Manila University. January, 1961. Mimeo.

109. Nigerian Government Commission on Post School Certificate and Higher Education. Investment in Education. (The Ashly Report). Lagos: 1960.

110. Philippine National Science Development Board, The Philippines First TV. College Course for Credits" Manila: August, 1961, Pamohlet.

111. The Public and the Programmes. Radio and TV Broadcasts to Schools and Colleges, 1964-65. London Brochure.

112. Rock, Vincent. "World-Wide Television and National Interest," Program of Policy Study in Science and

Technology. Washington, D.C., George Washington University, February, 1965.

113. Stanford University, Institute of Communication Research. Educational Television: The Next Ten Years. 1962.

114. UNESCO, Adult Education Groups and Audiovisual Techniques. Reports and Papers on Mass Communication. No. 25. Paris: 1958.

115. ----Developing Information Media in Africa. Reports and Papers on Mass Communication. No. 37, 1962.

116.___Developing Mass Media in Asia. Reports and Papers on Mass Communication. No. 37, 1962.

117. UNESCO, The Kinescope and Adult Education Reports and Papers on Mass Communication. No. 26, 1958.

118. ---- Mass Media in Developing Countries. A UNESCO Report to the United Nations. Reports and Papers on Mass Communication. No. 33, 1961.

119. ---- Screen Education, Teaching a Critical Approach to Cinema and Television. Reports and Papers on Mass Communication. No. 42, 1964.

120. Rural Television in Japan. A Report on an Experiment in Adult Education. Press, film, in the World Today. 1960.

121. ---- Social Education Through Television. An All India - UNESCO Pilot Project. Reports and Papers on Mass Communication. No. 38, 1963.

122. ---- Statistics on Radio and Television, 1950-1960. Statistic Reports and Studies. No. 8, 1963.

123. Television and Tele-club in Rural Communities. An Experiment in France. Reports and Papers on Mass Communication, No. 16, 1955.

124. ---- Space Communication and the Mass Media. Reports and Papers on Mass Communication. No. 41, 1963.

125. UNESCO, Visualaids in Fundamental Education and Community Development. Report on the UNESCO Regional Seminar in South and South East Asia held in New Delhi, India, 8-27 September 1958. 1959.

126. ---- Reports of the Meeting: Meeting of experts on development of information in Latin America, Santiago, Chile. 1-13 February, 1961. UNESCO/MC/41. Paris: May, 1961.

127. Television A World Survey. Reports on the Facilities of Mass Communication. Paris: 1952. Supplement 1955.

128. United Nations. International Survey of Social Development. 1955. E/CN.5/301/Rev.1- ST/SOA/21.

129. ---- Preliminary Report on the World Social Situation. 1952. E/CN.5/267. Rev. 1

130. ---- World Survey of Education. Paris: 1958.

Sponsoring Committee:

1. Professor Charles A. Siepmann
2. Professor Frederick L. Redefer
3. Professor Emilio L. Guerra

A Comparative Study of Educational Television in Selected Developing Countries and its Relevance to the Similar Use of Television in Afghanistan

HAFIZ SAHAR

Submitted in partial fulfillment of the requirements for the degree of Doctor of Philosophy in the School of Education of New York University.

1967

I hereby guarantee that no part of the dissertation which I have submitted for publication has been heretofore published and (or) copyrighted in the United States of America, except in the case of passages quoted from other published sources; that I am the sole author and proprietor of said dissertation; that the dissertation contains no matter which, if published, will be libelous or otherwise injurious, or infringe in any way the copyright of any other party; and that I will defend, indemnify and hold harmless New York University against all suits and proceedings which may be brought and against all claims which may be made against New York University by reason of the publication of said dissertation.

Sponsoring Committee:

1. Professor Charles A. Siepmann, Chairman
2. Professor Frederick L. Redefer
3. Professor Emilio L. Guerra

An Abstract of

A Comparative Study of Educational Television in Selected Developing Countries and its Relevance to the Similar Use of Television in Afghanistan.

HAFIZ SAHAR

Submitted in partial fulfillment of the requirements for the degree of Doctor of Philosophy in the School of Education of New York University.

1967

C. A. Siepmann
Chairman

Made in the USA
Monee, IL
04 November 2026

67386088R10111